Safety and Wearable Technology

Safety and Wearable Technology tackles the profound issue of workplace safety, as each day unacceptable numbers of workers lose their lives in accidents, with even more succumbing to work-related illnesses across the globe. Many safety professionals strive to reduce these risks, but they're often hindered by limited resources, time, and information to catch hazards before tragedy strikes. With wearable technology now readily available, a new frontier has opened for workplace safety, offering powerful tools that can change how we protect workers.

In this book, three experts in wearable technology explore the critical role wearables play in enhancing workplace safety. They walk the reader through the traditional challenges of safety management, making a strong case for the adoption of wearable technology. The book covers every essential aspect of implementing wearables in the workplace, from selecting the right technology and understanding technical approaches, to effective deployment and long-term safety management. It also addresses the leadership necessary to fully harness these advancements. Readers will gain insights into the innovations driving wearable technology, including sensors, machine learning, deep analytics, and artificial intelligence. As data-driven approaches redefine safety management, this book empowers safety professionals to gather actionable data, predict hazards, and intervene before incidents occur. This data-driven shift not only enhances safety but also redefines what the factory of the future can look like – a workplace where technology proactively protects workers.

Intended for professionals across safety, operations, and engineering fields, *Safety and Wearable Technology* will appeal to Health and Safety Engineers, Manufacturing Engineers, Plant Managers, Industrial Hygienists, Risk Managers, and many others looking to understand and implement wearable technology in their safety protocols. This guide is an indispensable resource for those ready to envision and create safer, smarter, and more connected workplaces.

Safety and Wearable Technology

Impact, Applications, and Implementation in Industry

Gabriel Glynn
Mark Frederick
Tom West

CRC Press
Taylor & Francis Group
Boca Raton London New York

CRC Press is an imprint of the
Taylor & Francis Group, an **informa** business

Designed cover image: image credited to Mākusafe.

First edition published 2025
by CRC Press
2385 NW Executive Center Drive, Suite 320, Boca Raton FL 33431

and by CRC Press
4 Park Square, Milton Park, Abingdon, Oxon, OX14 4RN

CRC Press is an imprint of Taylor & Francis Group, LLC

ISBN: 978-1-032-54988-0 (hbk)
ISBN: 978-1-032-55804-2 (pbk)
ISBN: 978-1-003-43230-2 (ebk)

DOI: 10.1201/9781003432302

Typeset in Times
by SPi Technologies India Pvt Ltd (Straive)

Contents

Part III Leadership, People, and Implementation, Tom West

Part V

Authors

Gabriel Glynn, COSS, CEO and Co-Founder, MākuSafe Corp., I grew up in a small trailer park on the edge of Cedar Rapids, Iowa, a typical Midwest town of about 125,000 people. My father worked as a machinist on 3rd shift at a factory in town, and my mother stayed home with four kids and often watched people with special needs to make extra income. My childhood summers were spent fishing in the nearby creeks and ponds and riding our bikes until the street lights came on. But even as a young child, I found myself fascinated with my father's work at the factory. I loved visiting and seeing the machines, hearing the sounds of grinding, pounding, and the clang of metal on metal. It was hard work, and I remember a couple trips to the hospital to pick him up after an accident at work. I didn't realize how impressionable those visits were on my life trajectory.

I was also fascinated with electronics and computers as a kid. I would take apart various electronics from around the house to repair them or turn them into other creations. By middle school, I had a job working for my neighbor's pawn shop where I would repair broken electronics, getting a cut of anything that was repaired and sold. We also had the first computer in our neighborhood, and I spent a lot of time learning how to build websites for local small businesses. By high school, I had my own tax ID, operating a small company that installed audio equipment and alarm systems in cars. Entrepreneurship was in my blood.

I met my wife, Amanda, while attending college at the University of Northern Iowa. We eventually got married and moved to Des Moines to start our life together. While she found work at a marketing agency, I really wanted to try my hand at building and growing a company. After starting a couple of small businesses and selling them off, I found myself with a great opportunity to join the ownership team at a small web tech company in the suburb of Ankeny. The company, Slash Web Studios, built websites and web software, with a focus on the industrial sector. I spent a great deal of time touring factories, finding ways to improve operations through web technology.

In 2015, we accepted a buyout offer and the search began for my next opportunity. At the time, my father had changed companies and taken on a safety leadership position at a large fulfillment center in Cedar Rapids. It was through spending time with my father that the idea for data-driven proactive safety was born. I took my knowledge of technology and applied it to an industry that was in dire need of a new approach. Thus, MākuSafe was born.

Amanda and I celebrated 18 years of marriage this spring, and we still live in Ankeny with our boys Caleb (13) and Tucker (10). MākuSafe has been an incredible journey that has taken me around the world to places like Australia, Singapore, Malaysia, Germany, France, Belgium, and the United Kingdom. I have met some of my favorite authors, like Dr. Erik Hollnagel, and visited corporate headquarters of some of the world's most iconic companies. It's been a blessing to be a part of such a passionate industry – workplace safety. There's still a lot of work left to do but the journey remains incredibly rewarding.

Mark Frederick, COSS, CTO and Co-Founder, MākuSafe Corp., Growing up in a small town in Minnesota, I had an obsession with taking my toys apart. Armed with screwdrivers, pliers, and sometimes a hammer, I would dismantle even my most cherished possessions. Anything with batteries, switches, or lightbulbs was fair game, and I marveled at the plastic parts, small circuit boards, and the occasional integrated circuit chip. I've always been a curious student of "things."

In 8th grade, the Sears store at our local mall had an Atari 800XL on display. While my parents shopped, I would lose myself in the computer aisle, devouring the programming manuals and writing BASIC code on the display model. Recognizing my passion, my parents gave me an Atari 800XL for Christmas in 1984. My life was never the same – I would stay up late most school nights, programming to my heart's content. I still treasure that computer today.

I pursued computer science in college and found my classes irresistible. I loved my coursework and often went above and beyond what was required. After graduation, my career naturally gravitated toward software development. I started by leading small teams and eventually grew into roles as a software architect, project manager, and thought leader focused on solving complex problems. Along the way, I spent eight years as a "road warrior," consulting for various companies across different geographies and technologies. Eventually, I traded my frequent flier status for more quality time with my family, landing at a Fortune 500 company where I worked on strategic initiatives in cloud computing, IoT, and artificial intelligence. I'm deeply thankful for my family's unwavering support throughout my endeavors.

These experiences laid the foundation for MākuSafe. A few years ago, I decided to push all my chips to the center of the table, embarking on this challenging journey with my co-authors Tom and Gabe to create something truly amazing. I've never looked back. If you ever find yourself with the chance to take a well-calculated risk that could change the course of your life, my advice is simple: do it.

Tom West, SPHR, SHRM-SCP, COSS, Vice President & Global Practice Leader, MākuSafe Corp., After our two co-founders at MākuSafe, I'm the next longest serving member of our team. I was an early investor and advisor to the company in 2016–2017, and then my friend Gabriel called. I'll never forget when he said, "I know it's Saturday, but do you think you can join the company full time on Monday?" I said yes without a hesitation.

Gabe and I had previously been friends for several years. We actually met while he was taking college classes more than 15 years ago and I was his instructor. Gabe was an exceptional student then, always asked great questions, approached things with a unique perspective, and was highly entrepreneurial. I had been an executive leader with many startup ventures, and Gabe later enlisted my help with several companies he was growing from time to time. I had no idea that one day I'd be working so closely at MākuSafe with the guy I mentored long ago.

I also felt fortunate that I finally got to meet and work with Mark, who possesses a number of remarkable traits. But his outstanding ability to conceptualize a solution we would take to the market and execute it so thoughtfully is one of the reasons we've achieved what we have. Together, Mark, Gabe, and I formed what Mark often referred to as the "needed three legs of our stool." Mark built our solution, Gabe provided vision and enlisted the required support, and I was tasked with developing our messaging, talking with industry, and building strategic and professional allies. We were developing a new market, educating company leaders about the potential for MākuSafe to make the world a safer place.

While Mark describes himself as a student of things, I have been a student of management and leadership development throughout my career. My experience is as a business owner and leader, senior human resources professional, trainer, consultant, and college teacher of 29 years. Throughout my career, I've always been focused on helping organizations operate more effectively and efficiently, enabling leaders to create more engaged, productive employees and strong safety cultures. I continue to be excited that this is such a big part of what we're doing here at MākuSafe. I am proud that over this time I have worked closely with and delivered educational presentations to the National Safety Council, ASSP, AIHA, and VPPPA, and been asked to be an Avetta Fellow.

From the outset we planned to co-author this book, with each of us intentionally focusing on our respective strengths. Gabe paints a vision for the future of wearable safety tech. Mark makes the complexities of technology easy to understand, eloquently shedding light on what's possible and why we developed our tech the way we did. My contribution is relaying what we've learned about the essentials for successful implementation, its impact on workers, and what this requires from leaders. We hope that it makes a difference in your operations so you can send more workers home safely every day.

Acknowledgments

To our peers, mentors, and colleagues in the safety industry: many of you laid the groundwork for our technology to fit directly into the hands of safety practitioners. Your advice, feedback, and guidance along the way has helped move our vision for a data-driven approach to safety into a reality.

To our clients from around the world: each of you faces unique challenges based on your type of work, geography, leadership style, and company culture. You helped us mold our vision into something that can work across all types of geographies and industries. Many of you were early adopters and without you we could not have gotten this technology into the field.

To our investors: without your financial support and belief in our team and mission, we would not have gotten a product built and into the marketplace. Your patience as we navigated setbacks, including a global pandemic, has driven our perseverance and success.

To our families: we're so grateful for your love and support. Late nights, long weeks, and months of travel around the world – you kept our spirits high when the road was most challenging. We could not have walked this journey without you.

To the MākuSafe team: your commitment to the mission, to supporting our clients, and to bringing new, innovative, ideas to the table has been what helps set us apart. We're not just a vendor to our clients, we're a partner, and it's because of your creativity and dedication.

To Jennie Morton: for your help in collecting our thoughts, stories, and ideas into this text. You are the glue that stuck it all together.

And finally to our readers: you picked up this book hoping to learn something about a new approach to safety. Thank you for taking this all in. We hope that it provides some practical and applicable insights into how you might approach safety differently.

Foreword

The origin story of MākuSafe has a poignant beginning – with an industrial disaster in 1919.

My great grandfather, John Griffin, was a factory worker in Cedar Rapids, IA. He was employed at Douglas Starch Works, which was a large-scale operation that produced cornstarch. However, the manufacturing process resulted in a fine and highly combustible dust that permeated the site. It was only a matter of time before an accident occurred. An explosion devastated the factory, and over 40 people tragically perished.

My great grandfather only lived to tell the tale because he left work over his lunch break to have a date with the woman who would become my great grandmother. It was by pure happenstance that his life was spared. My entire family trajectory could have been altered if he had kept his usual routine. It now seems like a divine calling that several of his descendants are part of the workplace safety industry.

My father, Brian Glynn, also had a long career in manufacturing, including as a machinist in a printing press factory, and later moved into management roles. In the early 2000's he accepted the role of safety director for a retail fulfillment center in Cedar Rapids, Iowa. One day over lunch, I asked him about the specifics of his job, and he made a joke about the amount of paperwork. Intrigued, I asked what he was documenting. His answer stunned me:

"I pretty much wait for something bad to happen, figure out why it happened, and then put a plan in place so it doesn't happen again. Then I go back to waiting for something else bad to happen."

In a nutshell, my dad had articulated the fundamental problem with workplace safety – it's reactive.

His comment left me wondering – can we use technology to help safety leaders like my father see risk before an accident happens, not after? So many companies rely on lagging indicators, yet an injury or death is the worst type of evidence that there was present. If safety leaders could identify hazardous conditions ahead of time, then there's a strong chance of intervening and mitigating them.

It was this simple premise that launched MākuSafe. Our idea is that when companies gather real-time data from wearable devices, they gain advanced knowledge of when and where risk is present. These red flags call attention to an area or a behavior that requires investigation. These alerts are what ultimately enables management to implement corrective actions that prevent a situation from creating an incident.

Up until recently, uncovering unknowns was the Holy Grail of workplace safety. But wearable sensors are a major turning point in our industry – we finally have the power to bring hidden risks to light. We're stepping into a new era where connected workers are the next evolution in Industry 4.0.

By Gabriel Glynn, founder and CEO of MākuSafe

About MākuSafe Technology

MākuSafe is an award-winning connected worker wearable technology and safety data analytics solution deployed across the globe. Our mission is to improve worker health, safety, and productivity by reducing injuries and mitigating workplace hazards and risk exposures.

The company was founded in 2016 based on a simple premise – because every individual worker's experience is unique to them, data gathered about environmental exposures and ergonomic motions from each individual is critical to understanding risk in the workplace. This data includes industrial hygiene and environmental concerns, slips/trips/falls, strain and exertion, ergonomic risk, and location/proximity.

This approach provides a more complete picture of workplace risk, trends and changes in those risks over time. It also integrates the worker into connected advanced manufacturing environments to accelerate industry 4.0 initiatives. In addition, the push-to-talk MyVoice capability improves communication by enabling real-time reporting from the frontline worker. Enhancing communication contributes positively to organizational culture, participation, and engagement.

MākuSafe offers a robust suite of comprehensive capabilities designed to provide real-time, individualized monitoring and notifications that enhance workplace safety and productivity. Below is an overview of the key features and benefits MākuSafe provides.

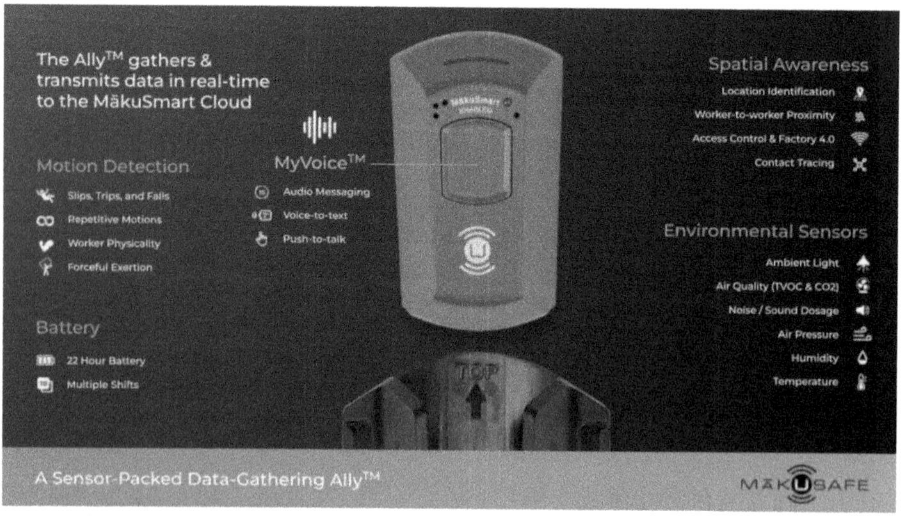

FIGURE 0.1 Strapped to the upper arm, the Ally wearable safety device carries a wide array of sensors and capabilities.

ERGONOMICS MONITORING AND PREVENTION

MākuSafe identifies and mitigates ergonomic concerns that can lead to injuries and reduce productivity. The system provides real-time alerts for potential issues such as slips, trips, and falls as well as identifying instances of strain and forceful exertion. By monitoring repetitive movements and overall physicality, MākuSafe helps prevent musculoskeletal disorders (MSDs) so workers remain healthy.

INDUSTRIAL HYGIENE MONITORING

Maintaining a safe and healthy work environment is crucial, and MākuSafe plays a vital role by offering comprehensive industrial hygiene monitoring. Whether workers are indoors or outdoors, the system tracks temperature, humidity, and heat index on an individual basis to prevent heat-related illnesses. It also includes a full noise dosimeter with seven-octave readings and DBA sound dosage (TWA) for hearing conservation programs. Additionally, MākuSafe monitors ambient light levels, which can impact worker fatigue. It also measures air quality to detect harmful gasses and fumes, including TVOC, CO2, and air pressure.

MYVOICE PUSH-TO-TALK COMMUNICATION

To enhance worker engagement and communication, the MyVoice push-to-talk feature allows workers to send voice messages by holding down a button and speaking into the device. Workers can report near-misses, share good catches, make observations, and conduct inspections with ease. The system also supports voice-to-text transcription, elevating worker participation by ensuring their perspectives are captured and communicated effectively. Audio messages can be delivered to dashboard users with optional notifications.

SPATIAL AWARENESS AND CONTACT TRACING

MākuSafe's spatial awareness capabilities provide detailed location identification by area, room, or workstation, which is overlaid on the facility floor plan. This feature enables the visualization of trends and patterns in real time, helping to identify areas that may require attention. The system also monitors worker-to-worker proximity, ensuring compliance with social distancing guidelines. In the event of a health concern, MākuSafe offers on-demand contact tracing to enable employer notifications, which is particularly relevant for COVID-19 infection prevention requirements.

COMMITMENT TO WORKER PRIVACY

Designed with a high degree of concern for worker privacy, the system does not collect any personal identifiable information (PII), ensuring compliance with HIPAA regulations. No biometric data is gathered, and workers are not subjected to haptic or other feedback. Furthermore, MākuSafe does not continuously track workers, maintaining their privacy and peace of mind.

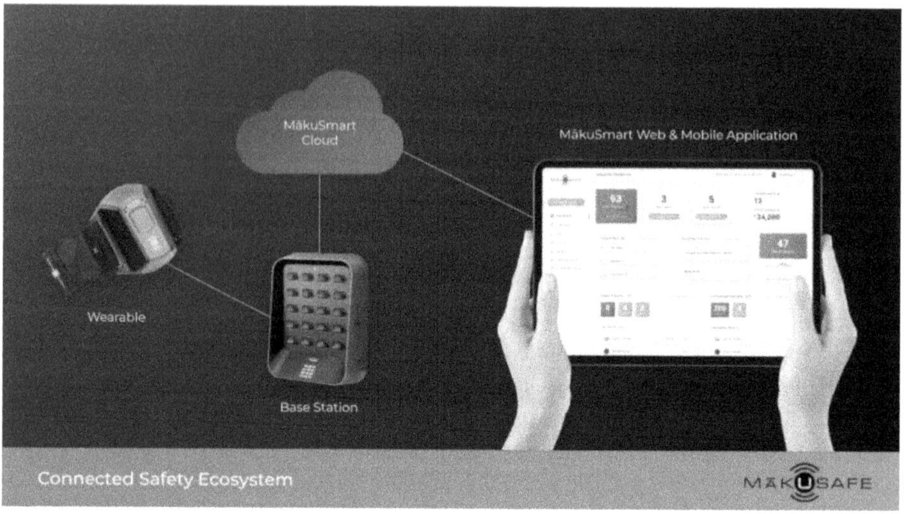

FIGURE 0.2 The Ally wearable device sends data in real time through the MākuSafe ecosystem to our cloud analytics platform. Within seconds, the latest risk and hazard intelligence and custom reporting are available to leaders in the MākuSmart dashboard. Each base station kiosk houses 20 wearable devices, and each device can be used by two different shifts per day.

MĀKUSAFE INTERACT AND INDUSTRY 4.0 INTEGRATION

As part of the Industry 4.0 revolution, MākuSafe Interact offers optional integration with other connected systems and Industrial Internet of Things (IIoT) equipment. This capability enhances productivity and efficiency by enabling worker presence and wearable proximity detection, further embedding the connected worker into the automated workplace.

SUPPORT FOR REMOTE AND FIELD WORKERS

MākuSafe extends its monitoring capabilities beyond the traditional workplace to remote and field workers. Wearables can be paired with mobile devices, providing GPS-enriched data via Android and iOS apps. This real-time feature allows organizations to gain insights into the experiences and exposures of workers outside the facility, ensuring comprehensive safety coverage.

MĀKUSMART: CLOUD SAFETY INTELLIGENCE AND ANALYTICS

The MākuSmart platform provides a powerful cloud-based safety intelligence and analytics solution. It delivers immediately actionable data insights on risks and hazards, allowing for at-a-glance understanding of the latest indicators, their locations, and who is affected. The platform's Location Explorer, SiteMap, and Trends Engine enable the study of data sets across timeframes and workgroups, providing a comprehensive view of safety performance. The system's open architecture supports integration with over 100 APIs and webhooks, ensuring seamless information exchange with other platforms and systems.

MākuSafe offers a comprehensive and integrated approach to workplace safety and productivity, leveraging real-time monitoring, advanced analytics, and robust communication tools to create a safer, more efficient work environment.

1 The Evolution of Safety Management – The Future Is Now

In the dimly lit control room of a manufacturing plant, the quiet hum of machinery was punctuated by the steady beeping of monitors. Claire Jackson, the plant's safety manager, sat at her desk reviewing the latest safety report. The numbers were stark. Despite decades of progress and a steady decline in injury rates, the past few years had seen little to no improvement. More alarmingly, fatalities were beginning to rise.

Claire knew these statistics all too well. The Bureau of Labor Statistics (BLS) data highlighted a troubling trend: While non-fatal Total Recordable Incident Rates (TRIRs) had plateaued since 2013, fatalities were climbing ("Fatal work injuries"). It was a pattern not confined to the United States; across the globe, "work-related deaths have jumped more than 12% since 2000, according to estimates from the International Labour Organization" ("Workplace deaths"). The question that haunted her and countless other safety professionals was simple yet profound: How do we break through this floor? How do we work toward zero?

The traditional approach to safety management has focused on lagging metrics such as incident rates and lost time. These are useful for tracking what had already happened but offer little insight into preventing future incidents. It is becoming clear that a paradigm shift is needed – one that emphasizes prediction and prevention over reaction. This is the dawn of a new era in safety thinking, often referred to as "Safety Differently" or "Safety 2.0."

FROM REACTIVE TO PROACTIVE: A NEW VIEW OF SAFETY

Lagging indicators have been the bedrock of safety management for decades. These metrics, while essential for understanding past performance, provide a rear-view mirror perspective. They told us where we had been but not where we are going. The stagnation of injury rates and the rise in fatalities are stark reminders of the limitations of this approach.

The leading causes of workplace injuries – overexertion, falls, and contact with objects or equipment – remain stubbornly consistent. These incidents are not just statistics; they are stories of human pain and loss. Claire reflected on the irony that most workplace incidents are preventable. The challenge is not in identifying what went wrong but in creating systems that prevent these incidents in the first place.

Leading organizations have long recognized the need for a proactive approach. The journey toward what we now call "Safety Differently" can be traced back to

DOI: 10.1201/9781003432302-1

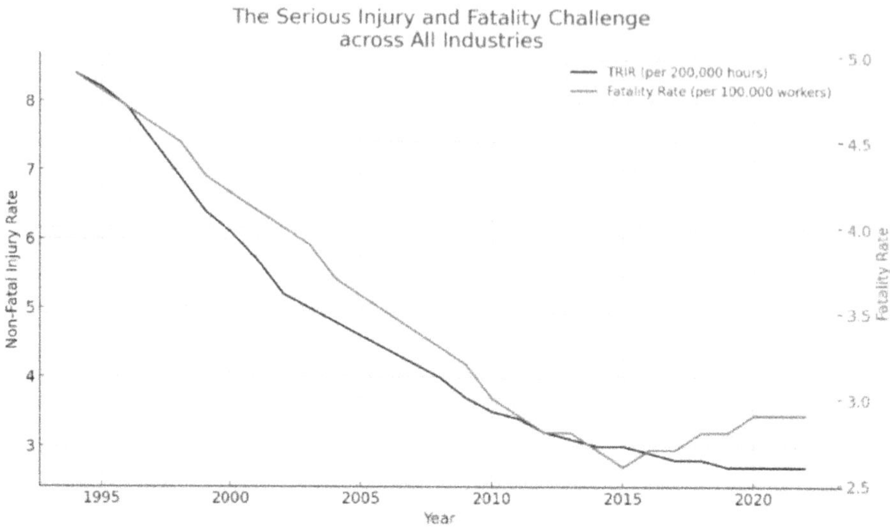

FIGURE 1.1 SIF rates have risen in the past decade, according to the Bureau of Labor Statistics ("Injuries, Illnesses, and Fatalities Databases").

pioneers like Paul O'Neill during his tenure at Alcoa. O'Neill's relentless focus on safety as a core value, rather than a priority that could shift, had set a powerful example. He understood that safety was not just about compliance but about culture and leadership.

Thought leaders, such as Todd Conklin, Thomas Krause, Sidney Dekker, Erik Hollnagel, Bob Edwards, Rob Fisher, and others, have since expanded on these ideas, advocating for a human and organizational performance (HOP) perspective. This approach recognizes that human error is inevitable and that systems must be designed to mitigate the consequences of these errors. It emphasizes the importance of learning and understanding work as it is actually performed, rather than as it is imagined in procedures and policies.

THE ROLE OF WEARABLE SAFETY TECHNOLOGY

As Claire continued to contemplate the future of safety management, her thoughts turned to the potential of emerging wearable safety technologies. The last decade has seen remarkable advancements in this area, with wearables offering transformative potential in preventing Serious Injuries and Fatalities (SIFs) as well as reducing overall incident and injury rates. Leveraging these technologies provides the predictive and preventative capabilities that traditional methods lack.

Predictive analytics, powered by data collected from wearable devices, offers a revolutionary approach to identifying potential hazards before they result in incidents. By analyzing real-time data on worker movements, environmental conditions, and physiological metrics, safety managers can identify patterns and trends that indicate increased risk. This data-driven approach enables more informed decision-making and proactive interventions.

Claire envisioned a future where every worker was equipped with wearables that monitors their vital signs and physical movements. Any anomalies, such as elevated heart rates or unsafe postures, would trigger alerts, allowing for immediate corrective action. This real-time data collection and analysis would be pivotal in preventing incidents before they occurred.

Artificial intelligence (AI) and machine learning (ML) would further enhance the capabilities of predictive analytics. These technologies shift through enormous datasets from wearables, identifying subtle correlations and predicting outcomes with remarkable accuracy. AI-powered safety systems could continuously learn from this new data set, improving their predictive capabilities over time. For instance, an AI system might analyze data from wearables to trend the likelihood of fatigue-related incidents. By understanding the precursors to SIFs, the system could recommend specific preventive measures, such as adjusting work schedules or providing additional breaks.

THE HUMAN ELEMENT: INTEGRATING TECHNOLOGY WITH CULTURE

While technology offers powerful tools for enhancing safety, Claire knew that its successful implementation depends on integrating these tools with a strong safety culture. Technology provides data and predictive insights, but it is the human element – frontline workers, safety managers, and organizational leadership – that ultimately determines how effectively these insights are acted upon.

Creating a culture where workers feel comfortable using new technologies and where their input is valued was crucial. Workers need to be trained not just in how to use new tools, but also in understanding their importance and potential impact on safety. Open communication and collaboration are essential to ensuring that technology complements human efforts rather than replacing them.

As Claire finished her review and prepared for the upcoming safety meeting, she felt a renewed sense of optimism. The integration of advanced wearable technologies with proactive safety management practices represents a powerful strategy for breaking through the current plateau in safety performance. By harnessing the predictive power of data, the practical applications of wearable devices and other new technologies, organizations can move closer to the elusive goal of zero incidents.

In the chapters that follow, we will delve deeper into the specific wearable technologies that are transforming safety management and examine case studies of organizations that have successfully implemented these innovations. By understanding both the potential and the challenges of these technologies, we can chart a path toward a safer future for all workers. The journey to zero is within reach, and technology is a key ally in achieving this vision.

Welcome to "Safety and Wearable Technology: Impact, Applications, and Implementation in Industry." The path to zero incidents starts here, with a commitment to see safety differently and leverage the full potential of wearable technology.

BIBLIOGRAPHY

"Fatal work injuries up in 2022." *Bureau of Labor Statistics*, 29 December 2023, https://www.bls.gov/opub/ted/2023/fatal-work-injuries-up-in-2022.htm

"Injuries, Illnesses, and Fatalities (IIF) Databases." *Bureau of Labor Statistics*, https://www.bls.gov/iif/data.htm

"Workplace deaths up 12% worldwide in past two decades: report." *Safety+Health*, 5 December 2023, https://www.safetyandhealthmagazine.com/articles/24808-workplace-deaths-up-12-worldwide-in-past-two-decades-report

Part I

Evolving Safety, Gabriel Glynn

2 Challenging the Safety Narrative

How pervasive are serious injuries in the workplace? The number of accidents and fatalities is eye-opening. Looking at the United States alone:

- There were 5,190 fatal work injuries recorded in 2021.
- This is an 8.9% increase over 2020, demonstrating a rising problem.
- For every 100,000 full-time workers, the fatal work injury rate was 3.6 fatalities, also up several points from the previous two years.
- These numbers mean that a worker dies every 101 minutes from a work-related injury on average, which is roughly 100 people per week.
- Progress on lowering these rates has been at a standstill for over a decade. Since 2011, the number of fatalities hovers around 5,000 annually ("National Census").

In Iowa, the headquarters of MākuSafe, the top sector is advanced manufacturing ("Advanced Manufacturing"). Yet it's shocking that the state averages about one worker per week lost to a fatality ("Fatal Work"). Nearly half are from preventable accidents at a jobsite, with leading causes from falls and exposure to harmful substances or environments ("Fatal Work"). Every death is a painful reminder that we can do better for our workers.

It's not only the loss of life that is incalculable. From a business perspective, there's also an undeniable financial toll on the economy. A fatal incident creates a ripple effect that impacts individual employers through downtime, lost production, decreased morale, and reputation damage. The likelihood of OSHA fines as well as litigation is also high. Not to mention every family who bears unexpected funeral expenses and outstanding medical bills while facing a future without a loved one's earnings.

It makes one wonder how far we've actually come in modern times. The Industrial Revolution was a major economic and societal shift, yet the advent of machinery also created treacherous working conditions. For example, if you tour a wool mill from the 1900s, you'll shake your head over the lack of safety precautions. Today's industrial facilities are a world apart in terms of protecting workers, yet risk management remains an ever-present concern. All businesses, from a family-owned factory to a Fortune 500 player, share the same challenge – how to minimize the danger and difficulty of the manufacturing process.

The usual advice to improve safety culture includes actions like clarifying expectations, evaluating processes, adding incentives, and following regulations. All of these are worthwhile efforts and will lead to positive impacts. But there's one surprising step that is often overlooked – an attitude check.

DOI: 10.1201/9781003432302-3

Many organizations have limiting mindsets about the origin of an accident. These entrenched beliefs act like blinders, obscuring the nuances of workplace risks. When these assumptions go unchallenged, they create a kind of tunnel vision that hampers safety efforts.

Let's dispel two of the most pervasive safety myths: 1) Work as planned translates perfectly into practice and 2) human error is the first place to lay blame after an accident.

WORK AS DESIGNED vs. DONE

"Work as planned vs. work in practice" is a phrase by Todd Conklin, senior advisor at the Los Alamos National Laboratory, in his book *The 5 Principles of Human Performance*. It acknowledges there's a gap between 1) how a company designed work to be performed and 2) how workers actually complete those tasks. Some variations are completely normal and expected. However, the space between expectations versus reality is where accident risk begins to sharply increase and move toward the failure line (see Figure 2.1). How does this happen?

1. "Work as planned" is disconnected from everyday conditions
   Just because something is designed a certain way doesn't mean that's the safest way.

   This is because work is often created in a vacuum. For example, the process was mapped by people who don't have first-hand experience on a factory floor. Or it was simulated in a virtual environment that didn't fully factor in a wide range of safety scenarios. Whether a human or computer

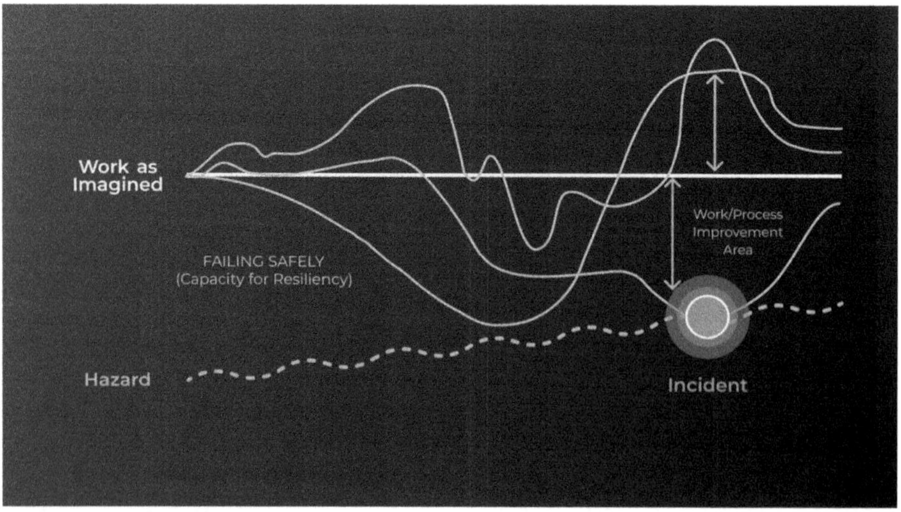

FIGURE 2.1 Work as performed, on good days and bad, does not follow a straight path as it was imagined or designed. Workers are forced to adapt to the circumstances of the day. By truly understanding the realities of how work is performed, we can improve work processes and create capacity to work safely, even when systems fail or people inevitably make errors.

envisions the work, it's extremely easy for the model to fail to predict minute changes that have a massive bearing on risk.

For example, the plan may not recognize that the manufacturing area is 95 degrees. This heat routinely causes workers' hands to be sweaty. In turn, that moisture affects their ability to grasp a lever. If their grip is compromised, their hand might slip. There are 11 levers where this creates no issues. But on lever 12, the wrong movement at the wrong moment will cause a ripple effect that escalates into a safety situation.

2. There's human behavior at play

   Even well-designed work doesn't perfectly match the reality of worker behavior.

   Work can – and often does – fluctuate from the design. This happens for any number of reasons, such as environmental stressors (unexpected glare), a lack of training (employees don't know any better or have forgotten), rushing to meet quotas (taking shortcuts), or variations in individual work styles (people have different body mechanics). Larger organizational issues such as being short staffed or increasing production expectations can also be at play.

3. The planned work doesn't anticipate unexpected changes

   The simplest alteration can have a domino effect.

   All it takes is for a single change to force a modification to the regular workflow. It might be an unexpected event like a machine breaking down or a larger situation like construction that reroutes the forklift path. While the change may look harmless on the surface, it could produce unintended consequences that heighten risk.

WHEN THE PLAN GOES AWRY

No matter their origin, all deviations create a dip away from the design's baseline. To be fair, not all dips are a blunder. Many don't even have the potential to cause an accident, much less represent an immediate safety threat. In fact, mistakes are so common that they should not only be anticipated but built into the process. Consider this argument by Conklin on his podcast *Pre-Accident Investigations.*

> "A worker making a mistake is super normal. Error is so normal, it's not even interesting. And it's never causal. Error is just an operational phenomenon that happens all the time. The reason it can't be causal is because error doesn't always create failure conditions. Workers screw up all the time, but not every screwup has high consequences of failure. Most of them, you can catch or correct, or they just don't matter. That sets a weird precedent … that error is normal. The battle isn't taking error out of a system – it's creating a system that can tolerate the normal error that's always in that system"
>
> (Conklin, "Safety Moment").

It's when error hasn't been factored into the process that variations start to compromise safety. Once these mistakes gain momentum, they predictably move closer to the failure line – the point of no return when an error results in an accident, injury, or fatality.

This is where many organizations have a blind spot – *they have no idea how many times they've brushed the edge of that failure line.*

No accidents, no problems, right? The assumption is that a lack of incidents means that work is being done as described and designed. But what the company doesn't realize is that they had skirted the failure line a hundred times before a SIF finally happened. They were always on track for this accident because no one was looking for dips, much less measuring their impact on safety.

> "If a facility has vulnerabilities for SIFs and is not paying attention to those precursors in a systematic way, focusing on small injuries is counterproductive. It allows SIF rates to stagnate and employees to get hurt. It leaves the most senior leaders flummoxed when a serious injury occurs seemingly out of the blue – while the worker who sees the vulnerabilities day in and day out counts the event as inevitable,"

Argue Krause and Bell, two organizational safety experts, in their book *7 Insights into Safety Leadership* (27).

How do we address this mindset? Companies need to routinely evaluate their safety culture to pinpoint where risk has been overlooked. That means asking questions like:

- Where are things going well?
- Where are there departures from work as designed?
- Where are these variations creating near misses?

Each of these should be followed up with the simple question of "why?" If an area has a stellar safety record, what is driving that? If workers are deviating from the planned work, what's the reason? Knowing is better than assuming. When an organization isolates good safety practices, they can iterate those precautions throughout a site or even across multiple locations.

Once organizations start analyzing precursors, they often stumble across a surprising truth – few accidents are caused by a single failure. Tragic incidents are frequently the result of a chain of events that was influenced by business decisions and organizational behavior. Remember the phrase "when you point one finger, there are three fingers pointing back at you?" Preventing a SIF means taking a hard look in the proverbial mirror and reflecting on how the entire company contributed to safety risk.

THE WORKER ISN'T ALWAYS THE ROOT CAUSE

The root cause mentality assumes there was a single point of failure that occurred at a single point in time and was triggered by a single individual. This default reaction places the blame entirely on the worker. If they did something different than how work was designed, it is simply evidence of their neglect.

Instead, companies need to flip the script and work backwards from an incident. What were all of the decision points that led to this one moment in time? There were likely dozens – if not hundreds – of factors that increased risk until the SIF's tipping point. They often originate from operational changes that trickled down from the corporate level, an unexpected repair, or external complications like supply chain delays.

Let's run a hypothetical SIF that resulted in the death of Jim. A facility has a large robotic arm that picks up steel plates and places them on the CNC machine to be laser cut. The equipment is protected by a locked cage. Furthermore, employees are trained to never enter this area while the machine is running. Jim is an experienced worker and has been at this station for over a year. But for unknown reasons, Jim bypassed the safety gate on the machine and went inside the cage only to be crushed to death.

Root Cause Reaction

The conclusion – "Jim was an idiot."

It's unfortunate that a common refrain in the industry is "you can't fix stupid." The problem is that when one person is declared the root cause, the company feels absolved from examining any circumstances that preceded that event. All of the responsibility is placed on the employee who was hurt or died, and it's back to business as usual. The accident is categorized as an outlier rather than the tip of an iceberg.

End result: The company sees the incident as the result of individual behavior and takes no further action.

Domino Effect Reaction

The conclusion – "We accepted a certain amount of risk, and Jim's the victim of our poor planning and decision making."

The reality is a SIF is rarely the result of a single point of failure. A company that prioritizes world-class safety will use a SIF to peel back the layers. They will start asking questions like:

- Why did Jim do that? What was he trying to accomplish?
- Why did he feel like he had to take this risk? Was he under pressure from other areas to meet deadlines or goals? Was this action perceived as an acceptable shortcut?
- Was this an isolated incident or not? Or were there 500 other times that Jim was able to get in and out through the gate without harm? Did this lead to a false sense of security?
- Was there a machine malfunction? Is there always a part that gets stuck or needs adjusting that would cause someone like Jim to enter the gate?
- Have we created a lax environment that makes people comfortable with bypassing safety protocols?

The company might find that a series of interconnected influences created the perfect storm for this failure. For example, the machine guard broke the previous month. Even though Jim put in a work order, he was told to continue working without it.

This was because corporate made a decision the year earlier that the only way they could bonus people was based on productivity. So they cut back on expenses, which meant they didn't have a secondary machine or spare parts on standby. Meanwhile, maintenance acted on Jim's work order, but it was going to take two months for the new part to ship from overseas. That left at least a month where Jim was working without the proper protection. It was in that window that Jim made a decision that cost his life.

End result: The company sees the incident as the result of risky organizational behavior and forms a plan to correct course.

KEY TAKEAWAYS

- Don't assume work as designed is how it's being done
- Don't assume a lack of accidents means a lack of problems
- Don't assume the worker is the root cause of an accident

BIBLIOGRAPHY

"Advanced Manufacturing: Where Manufacturing Keeps Advancing." *Iowa Economic Development*, 2024, https://www.iowaeda.com/advanced-manufacturing/

Conklin, Todd. *The 5 Principles of Human Performance: A Contemporary Update of the Building Blocks of Human Performance for the New View of Safety*. 2019.

Conklin, Todd. "Safety Moment – Principle Number 1 – Mistakes are normal." *Pre-Accident Investigations*, 2 May, 2018, https://preaccidentpodcast.podbean.com/e/safety-moment-principle-number-1-mistakes-are-normal/

"Fatal Work Injuries in Iowa – 2022." *U.S. Bureau of Labor Statistics*, 7 March 2024, https://www.bls.gov/regions/midwest/news-release/fatalworkinjuries_iowa.htm. Press release

Krause, Thomas R. and Kristen J. Bell. *7 Insights into Safety Leadership*. The Safety Leadership Institute, 2015.

"National Census of Fatal Occupational Injuries in 2022." *U.S. Bureau of Labor Statistics*, 19 December 2023, https://www.bls.gov/news.release/pdf/cfoi.pdf. Press release.

3 Data-Driven Businesses Include Safety

When your organization talks about data-driven decisions, is safety included?

Businesses are using data to inform every area of their operations in ways that would have been unimaginable even a decade ago. Because of real-time insights, companies can calculate everything from the timing of layoffs to investing in capital improvements. They are digging into the nuances of who their consumers are and their buying patterns, then translating those factors into product positioning. They are doing the same profiling on purchasers so they have more negotiation power. They can even monitor daily price fluctuations across the supply chain. Up and down operations, leaders are getting more data savvy.

But safety is left out of the discussion in many cases. In fact, it's often one of the last business units to undergo digital transformation. There are three culprits that keep workplaces from fully embracing the capabilities of safety data.

#1 – DECISIONS ARE DRIVEN BY INTUITION

There are few instances where a business would lean entirely on gut instinct. No leader is going to use wishful thinking to determine if it's going to be a banner year or if they need to order more components to fulfill orders. They're going to approach those considerations armed with verifiable evidence and honed forecasting. Safety decisions should also be founded in statistical data – otherwise blind optimism is in the driver's seat and could lead you astray.

For example, a business wants to increase productivity and revenue, so they add robotic welders to boost output. Their first oversight is assuming the robots won't create any complications, which means they didn't consider the downstream impact on shipping and receiving. Second, they are not gathering any critical data that is useful in understanding the change, like environmental and ergonomic sensors that could alert them to trouble. The reality is that production is outpacing the warehouse, and this has caused an unintended bottleneck that is escalating into a safety issue.

First, product starts piling up in different areas. Pallets are stacked so high that they block out the lighting and reduce visibility. The overflow also causes the team to pack pallets in spaces not intended for this activity, which is problematic because they now overlap with where the forklifts drive. The consequence is the forklift routes are detouring unexpectedly. The combination of decreased light and modified forklift paths is a perfect storm for someone to be seriously injured or killed.

What happened is that new risk was introduced because an operational change was made in a data vacuum. This is a failure twice over because the delays in shipping and receiving cancel out the welding department's productivity as well as significantly increases the likelihood someone will get seriously injured or killed.

DOI: 10.1201/9781003432302-4

What could have halted this drift toward the failure line? First, physical exertion data from workers would make it noticeable when there's a point of departure. Second, mobile sensors that detect ambient light could flag when it has fallen below acceptable levels. If both of these had been treated as leading indicators, the company could have called for a safety investigation.

#2 – ACCIDENT DATA IS THE SOLE INDICATOR OF RISK

Traditionally, many bad things must happen in order to have enough data to identify a safety trend. The paradox is that the more accidents you have, the more data there is to trend. Except that's an incredibly poor way to get data. Businesses shouldn't pass a certain threshold of injuries before taking action. It would be like waiting to plan for workplace violence until a fight broke out – it's a passive approach that doesn't make any sense.

Accident rates also shouldn't be used as the sole predictor of future behavior. Keep in mind SIFs and DARTs exclusively look at past events. SIF reports can also be unreliable in trying to understand what truly happened because other critical factors are often omitted, such as environmental factors, fatigue data, and production pressure. Organizations would have a much better understanding of where risk is if they used near miss (good catch) data and leading indicators rather than relying on multiple SIF reports.

#3 – ONLY STATIC SENSORS MONITOR SAFETY

Smart factories frequently deploy sensors to detect hazardous conditions. However, these devices have limitations because they are fixed to one location and only detect within a set radius. A more advanced and precise method to gather data is to deploy wearable sensors on workers that gather data in real time, giving a much clearer picture of what the individual worker is experiencing.

Consider a wall-mounted temperature sensor. While it generates a good set of data, it's specific to what the wall is experiencing. It can't tell that a person who is 20 feet away is standing next to a machine that's producing a dangerous level of heat. The worker's experience is totally different than the wall reading, even though they are in the same room. This leads to an awareness gap because the stationary sensor doesn't have adequate proximity to a heat source to monitor risk.

Now imagine the worker is wearing a temperature sensor that generates alerts when heat exceeds safe levels. The picture changes completely. Employees could protect themselves by taking breaks and hydrating. They might even stop to inspect the machine for performance concerns.

SAFETY DATA ELIMINATES TRIAL AND ERROR

Risk data takes the adage of "you can't manage what you don't measure" and puts some real teeth behind it. By eliminating guesswork, safety data empowers businesses to make decisions based on evidence. They can:

- Eliminate assumptions about working conditions
- Weigh options that will lower risk

FIGURE 3.1 As the hot equipment example illustrates, risk can be subtle. It can also be cumulative, building over time. Just because accidents aren't happening doesn't mean that work is being done safely or environmental hazards aren't present. It's why it's so transformative to have employees generating safety data regardless of their location. They help an organization identify where work is deviating from the design, prompting a timeout to investigate. It's this ability to stop, assess, and correct that prevents dips from toeing the failure line.

- Compare safety gains against investment costs
- Prove that a work redesign achieved safety goals

Quantifying the impact of a change is where safety data is particularly powerful. Implement too many modifications at once and it is nearly impossible to isolate which one moved the safety needle. It's like making a diet switch by adding and subtracting multiple foods at the same time – there's no way to determine an improvement was the result of one particular change.

But with a baseline of safety data, a business can create control groups for evaluation purposes. They can implement one option and measure the impact against the baseline, do the same for a second option, then compare both outcomes to each other. To determine which course of action is the best, they can then proceed to a cost-benefit analysis and use real data to support it.

CONVEYOR LOADER vs. NEW HIRE

Let's revisit the case of adding robotic welders to increase productivity. In this version, all employees have a wearable safety device that captures motion data. The plant has an established record of how much physical effort is typical for each department. They also monitor for fluctuations that indicate workers are expending more energy than usual.

The ripple effect from the robotic welders is that the two-person dock crew is struggling to keep pace. This is documented as a sharp rise in physical exertion over a two-week period. The change in physical exertion is classified as a leading indicator because it shows increased potential for overexertion, strain, repetitive motion, and injuries. The conclusion is that ergonomics is being compromised in the shipping and receiving department. Since that is not an acceptable risk, corrective action is warranted.

Rather than wait for the loaders to complain or get hurt, leadership proactively asks the affected team for their viewpoint and starts an informed discussion about how to resolve overexertion. Through their discussions, two solutions are proposed: 1) add a conveyor for loading and unloading trucks or 2) hire a third person.

To verify the safety gains associated with either, the factory tests each solution individually and updates the motion data for the new conditions. Now they have a before-and-after picture of exertion for both choices. This data allows the company to balance safety with other factors like cost and long-term planning.

Conveyor	Third Loader
Reduces physical expenditure by 50%	Reduces physical expenditure by 70%
One-time investment cost + associated upkeep	Ongoing compensation + benefits but no upfront costs
Less expensive but less effective	More expensive but more effective

Leadership can then embark on sophisticated decision-making that is free from conjecture. If lower cost is a priority, then the conveyor belt is a better option because it will decrease risk by an acceptable margin while satisfying shipment demands. If cash flow is a concern, then hiring a temporary worker will be prudent until the company can invest in equipment down the road. If the labor market is tight and staffing is difficult, then the upfront costs of equipment are more justifiable. It's also plausible the factory will adopt both measures. This might create an exciting opening to add another robotic welder because there's proof that higher output won't sacrifice safety.

All of these considerations are possible because they are rooted in one thing – an IoT connected worker generated data associated with a leading indicator. The best part? The entire decision process was done before there was ever an accident or injury. This is just one of the many ways connected workers and safety data are part of continuous improvement.

KEY TAKEAWAYS

- Operational changes made without data often increase risk
- Accident data alone doesn't give a full picture of risk
- Safety data eliminates guesswork

4 Green Flags in Safety Culture

How likely are your workers to come forward with a safety concern?

Across virtually every organization we talk to, there are few companies that are good at gathering safety information from workers. Communication barriers – perceived or founded – are a significant impairment to collecting critical feedback from the factory floor. Any employee hesitation often boils down to three uncertainties – is it convenient, safe, and worthwhile to report an issue?

First, recognize if there are real-time pressures on employees that take priority over reporting an issue. Stepping away from their workstation for even a few minutes to flag a hazard can have immediate and negative consequences, such as impact their ability to reach quotas, affect other people down the line, or disrupt the entire assembly process. These impacts are a disincentive for them to alert a supervisor about lagging or leading indicators, much less near misses.

Second, the reporting process is often manual and cumbersome. Whether paper or electronic, it takes time to fill out documentation. If the reporting tool is confusing, too complicated, time-consuming, or overly generic, employees won't bother with it.

But the biggest challenge often lies in how empowered employees are to say something in the first place. Leadership's receptiveness to feedback, validation of concerns, and willingness to implement a solution can make or break this aspect of safety culture.

EMPOWERING WORKERS

Consider that the number one fear that prevents someone from speaking up is retaliation. It's like company surveys that claim to be anonymous but clearly are not – no one is fooled. Employees shouldn't be worried that there will be repercussions if they shine a light on a problem. If everybody's afraid to say something, no amount of PPE and advanced reporting tools will reverse the damage. Staff need to see proof that they can freely go to management if they see something unsafe or have a suggestion to reduce risk. They need to know that concerns will not only be protected but celebrated for their willingness to come forward.

There's also nothing more damaging to safety culture than somebody taking the initiative to speak up and then seeing nothing is being done. All it takes is for one person to be ignored and it breeds a perception that the company doesn't care. This is especially toxic if a number of people have raised a concern and management never acts on the information. This also happens when a report is submitted but sits

DOI: 10.1201/9781003432302-5

there for hours with no review or action, to the point where the information is now completely useless due to the delay. You can't blame employees for beginning to think, "I don't know why I say anything because nothing ever gets resolves. No one takes us seriously so I might as well keep quiet and just do my job."

A lack of communication can also unintentionally compound this problem. It's possible there's concrete progress at the leadership level that employees can't see. Management could be having meetings about how to solve an issue or working on the budget to purchase a solution. Maybe they even went ahead and ordered equipment that will reduce risk but are waiting on a two-month shipping delay. But if no one is communicating updates to the floor staff? It still has every appearance that safety is only given lip service.

Trust and transparency are paramount to any safety culture. Without accountability and follow-through, safety data can only be a benchwarmer rather than a most valuable player. One of the most simple yet effective tools we have seen our clients use for transparency is a "You Said, We Did" board. One side has comments and suggestions from frontline employees (You Said), the other side is what has been or is being done (We Did). This is done on a rolling basis, even if resolution is still in progress. For example, the "We Did" update might be as simple as "We took this to our weekly management meeting and discussed it. It's on the agenda for next month to see if this is still a concern." That kind of transparency shows that not only is everyone in the organization being heard, but leadership is making a concrete effort to rectify issues as well.

HOLISTIC SAFETY LEADERSHIP FROM L'ORÉAL

Our team had the privilege of attending a presentation by Malcolm Staves, Global VP of Health & Safety for L'Oréal. With 88,000 employees and over 40 manufacturing facilities, safety is mission critical to this beauty leader ("About L'Oréal"). We were thrilled to learn how L'Oréal prioritizes leading indicators as well as holistic safety approaches. They even go as far to ask a startling but revealing question at every location:

"How are we going to kill somebody with what we're doing?"

This is how they evaluate every square foot of a facility. No location is exempt from this process – even zones where no one has ever been hurt doing a particular job or locations with low incident rates are scrutinized with this question in mind. It's how they force themselves to uncover and understand leading indicators, finding those essential lessons learned that can be applied companywide.

They also include their workforce as "active agents" in their safety culture, "using their acuity and skills to be part of the protection process" ("L'Oréal, A World-Class"). Rather than mandates to passively follow, "L'Oréal is keen to instil [*sic*] in employees a sense that their individual contributions to creating and maintaining a safe workplace, however small, are vital in creating a self-sustaining bottom-up (rather than top-down) safety ethos" ("L'Oréal, A World-Class").

What's even more impressive is how L'Oréal has framed safety as a necessity that extends beyond working hours, encouraging staff to be health and safety ambassadors in their personal lives. As Staves explained in an interview with *Safety & Health Practitioner*:

> "We do not expect anybody to work safely only for L'Oréal – then they are potentially working safely to avoid being a statistic. We have changed the conversation dynamics to the why and the sense of purpose and the need to work safely for themselves and for their families so they can spend time outside of work being with the people they love doing what they love to do" ("'It's now all").

LOOK FOR THE GREEN FLAGS

How are you documenting leading indicators?

Leading indicators are one of the most beneficial metrics in a safety program. They come in many forms, from safety meetings and refresher training to proactive maintenance and PPE compliance. It's an amazing shift in safety culture when an organization moves from responding to injuries (lagging indicators) to proactively mitigating potential hazards (leading indicators).

But we'd like to suggest another type of leading indicator to investigate – areas of no accidents.

Safety professionals are so attuned to red flags that we've forgotten there are green flags too. The absence of data is a compelling story that demands attention. When a zone or a team has a low incident record, it's a strong sign of a healthy safety culture. You want to investigate what a department did or is doing that is positively impacting safety.

This is where data can help you conduct a positive form of root cause analysis – uncovering a safety solution that's already a success in your manufacturing environment. There are two data sets that will reveal best practices that have been flying under the radar: motion data and accident-free employees.

Motion Data

Measuring physicality offers an incredible wealth of insights. Take a factory with a hundred workers doing the same job requiring the same output. Due to differences in height, strength, and body mechanics, it's normal for physicality data to show a range in exertion. Companies tend to focus on workers who are at the top of the bell curve. They often assume – and rightly so – that individuals exerting more energy are either not working efficiently, at risk of an injury, or both.

But they also need to pay attention to the lowest risk workers. There is a reason they are at the bottom of the distribution – it's not pure luck or statistical likelihood. Chances are they've done something intentionally or even subconsciously to improve safety. These employees are learning opportunities.

For example, an employee welded a custom mechanical bar onto their workstation so that they can flip a heavy piece of material over with greater ease and speed. Because they are not having to physically maneuver the part every single time, they've accomplished two safety improvements: lowered injury risk from repetition motion and strengthened ergonomics.

But it would be easy to overlook this type of self-directed safety improvement. No one asked them to make this modification – they simply innovated on their own. There's a small chance the solution will appear on a safety inspection or be mentioned during a briefing. However, the worker or their supervisor isn't thinking about the bigger safety picture – it's not their job to consider how this single improvement could be scalable throughout the plant and at what cost.

So what would bring leadership to this individual in the first place? Motion data. This worker's low physicality ranking would stand out among colleagues. That's a green flag begging to be investigated. Safety managers can connect with this individual to learn what's contributing to their score, keeping an eye out for replicable solutions. Even if the safety solution or habit is one-off and unique to this person, there's benefit in recognizing their efforts and highlighting them as a role model.

Injury-Free Workers

Far too often, safety investigations are only focused on people who got hurt. If your only measurement of safety success is the rate that people are still getting injured, that's a short-sighted way to assess risk.

What if we looked at people who have no history of accidents?

This turns the safety model on its head. Because if you have 300 employees who are free of incidents, their track record is another type of green flag. The data is telling us this worker pool is actively supporting the safety culture – that's worthy of investigation. Their actions or habits are an untapped source of constructive insights.

REAL-TIME REPORTING

What if workers were wearing a device that allowed them to create a quick recording that was immediately sent to the attention of safety leaders?

Our voice reporting function does exactly that – bypassing paper forms, phone apps, and electronic logs. It takes mere seconds to make a recording with the same ease as a voicemail. It can be done in real time, no waiting to remember an issue at the end of the shift or hunt down a supervisor before heading home. The information goes directly and quickly to designated leadership, allowing them to address immediate hazards or start conversations about areas of concern.

This feature is one method to overcome the lack of near-miss (good catch) data, when the failure line was within sight but narrowly avoided through good fortune. It's also a way for workers to note when work is drifting away from the design before there's a situation heading for trouble.

MākuSafe's platform logs the voice recording with a speech-to-text transcription. Additional documentation such as photos, detailed notes, updates after an investigation, and mitigation actions can all be included. This creates a robust and referenceable entry stemming from the initial communication. There is even an option to create a word cloud with commonly used terms from the messages to visually depict what employees are reporting.

One of our global clients uses the voice function to gather information about near misses. Using behavior-based safety observations, employees are asked to spend a few minutes of their day anonymously monitoring a peer while working. Rather than a punitive "gotcha" exercise, the purpose of the observations is to have an objective set of eyes spotting hidden issues. For example, it might take the viewpoint of a coworker to notice that someone is improperly rotating their body while moving materials. The motion is likely subconscious rather than a flagrant safety violation, but it still needs to be treated as a leading indicator because of the likelihood of strain or a repetitive motion injury.

Voice reporting is also a meaningful way to highlight where someone is contributing to safety culture. These are the "good catches" that need to be rewarded. They are anything from proactive behaviors and adherence to policies, to corrective action that reduces workplace hazards. For example, Sam is observed consistently using multiple points of contact while lifting parts – this is conduct that should be recognized and positively reinforced. Voice reporting is an efficient and easy way to report good catches, ultimately contributing to the strength and efficacy of your safety culture.

KEY TAKEAWAYS

- There are many barriers that prevent employees from reporting safety issues
- Trust and transparency are the foundation to any safety culture
- Documenting leading indicators ("green flags") is essential

BIBLIOGRAPHY

"About L'Oréal." *L'Oréal Groupe*, n.d. https://www.loreal.com/en/group/about-loreal/

"'It's now all about putting people back on the right track': SHP speaks to Malcom Staves, Global VP of HS at L'Oréal." *Informa Markets*, 2024, https://www.shponline.co.uk/culture-and-behaviours/its-now-all-about-putting-people-back-on-the-right-track-shp-speaks-to-malcom-staves-global-vp-of-hs-at-loreal/

"L'Oréal, A World-Class Safety System." *L'Oréal Groupe*, n.d. https://s32891.pcdn.co/wp-content/uploads/2023/07/Loreal-A-World-Class-System-digital.pdf

Part II

The Technology, Mark Frederick

5 Why Wearables Aren't a Fad

Let's clear up one of the biggest misconceptions about wearables – they are not a craze.

Considering that the earliest versions arrived in the 12th century, we can't even call them a trend. They predate the arrival of computers, cars, electricity, and even steam power by hundreds of years. Wearables are a well-founded part of the human experience, fundamentally changing societies with every iteration. Wearables are an unstoppable force.

Supported by advancements in consumer electronics, wearables are steadily moving into the workplace. It's not only because the incoming generation is already comfortable with high tech – our "always connected" mindset has normalized wearables across most age groups. In 2022, wearable users were estimated at 1 billion based on sales figures. We're only going to continue down this path – the market share was $27.91 billion in 2019, more than doubled to $61.30 billion in 2022, and is projected to hit $74 billion by 2025 (Beckman).

This widespread adoption is paving the way for wearable safety technology to become a regular, everyday tool. From exoskeletons to adhesive patches, there's a wide aperture on how wearables already are and will be standard on a job site. It's only a matter of time before they become as commonplace and essential as phones and computers. It's a confident projection that more workers will be equipped with wearable technology that helps them do their job not only more efficiently but more safely.

DEFINING A WEARABLE

Wearable technology has two defining qualities. The first is that it is worn on a human. While it may seem like an obvious statement, we can't overlook the historical context of how we got to where we are today. Many mechanical engineering and computing capabilities had to be invented to arrive at this point. Along the way, we have been conditioned to accept assistance from technology on or in our bodies for centuries. In fact, you could argue that the human body has always been a target platform for technology in one way or another.

The second characteristic is that the wearable is participating in an information exchange. It's either a consumer of or contributor to information; sometimes even both in the same device. No matter how it's generated, this data is used to enhance decision-making.

In every version of wearable technology, there's an undercurrent that remains constant – our desire to be autonomous. Wearables provide a certain level of independence because we have essential information directly on our person. Humans have

DOI: 10.1201/9781003432302-7

long benefited from these insights, from the simple ability to check the time or read words more clearly to taking measurements of our body's most vital functions. We're so comfortable with wearables that we've even allowed medical devices to be implanted inside our bodies. We are living more healthy, productive, and satisfying lives because of wearables.

Safety wearables have a long lineage of distinguished forerunners. Each of these categories – cultural disruptors, life savers, personal entertainment, and mobile communication – has a direct influence on our technological ability and societal willingness to wear a smart device at work.

Wearables as Cultural Disruptors

One of the earliest examples of wearable technology is optometry, which debuted around the start of the 1300s. People take it for granted these days, but it required a significant leap in glass technology to adapt and refine handheld lenses for the human body. It was revolutionary when people could wear magnification to see their world more clearly.

Another wearable milestone was the pocket watch in the early 1500s ("A Guide to History of Pocket Watches"). Prior to that point, telling time was a public service in the form of clock towers. Technology that once loomed over a town square was eventually miniaturized so it could hang from a small chain. The first known wristwatch then landed on the scene in 1810, designed for the Queen of Naples ("1810, The First Wrist Watch"). Once wearable timepieces graduated from a symbol of wealth and were available to the everyday consumer, tracking hours and minutes was democratized.

It's not hard to imagine naysayers in those early days who claimed watches were "just a trend" and "people would soon get tired of having them on their wrists." Yet look at where we are today – you'd be hard-pressed to find someone who isn't carrying a clock on them in some form or another during waking hours. Once the time went on our bodies, we never took it off.

Wearables as Life Savers

Another revolutionary moment in wearable development was the introduction of medical devices. These breakthroughs came with the invention of the transistor in 1947 by Bell Laboratories ("1952: Transistorized"). Technologies that once relied on vacuum tubes could be scaled down into lightweight devices. For example, the first commercial hearing aids worn completely in the ear were released in 1952 from Sonotone ("Inventing the Transistor").

The introduction of the battery-operated pacemaker was a major turning point in healthcare. Medtronic is credited for the first wearable external device in 1957, which adapted pacemakers that had to be plugged in and housed on a cart during open-heart surgeries; they produced their first transplantable model in 1958 ("75th anniversary"). Wilson Greatbatch, an engineer working under the VA hospital system, holds the patent for the first self-contained, fully implantable pacemaker in 1962 ("Wilson Greatbatch"; "The Invention of the Cardiac Pacemaker"). It is no small thing to imagine how many people's lives have been extended thanks to a wearable on one of our most important organs.

This wearable springboards into implantable cardioverter defibrillators (ICDs), which were successfully demonstrated during clinical trials throughout the 1980s and refined to terminate two types of abnormal heart rhythms by the early 1990s (Van Welsenes et al.). It's amazing to think that people at risk of cardiac arrest now have the reassurance that a heart problem is being mitigated by a piece of technology that's doing a better job than their own body.

Even though pacemakers and defibrillators are a closed loop system, they still demonstrate the power of a responsive wearable. They collect information about how our bodies are functioning and can automatically generate a corrective action to alleviate a problem that could lead to death otherwise.

The 2010s and 2020s have also been incredible decades for fitness and health wearables. Today we have multiple brands of smart rings that measure blood pressure and heart rate as proxies for mood and focus. The *Inspire* implant relaxes airway muscles to address sleep apnea as an alternative to CPAP machines and even has an accompanying app that tracks data (*Inspire*). Diabetics can opt for any number of real-time glucose monitors with a sensor on or even under the skin. There have even been inspiring advances in bionic prosthetics equipped with artificial intelligence, allowing people with amputations to feel sensations and control movement that better mimic a natural limb. We are living in a moment of increasingly personalized wellness data – all generated from wearables.

Wearables as Personal Entertainment

Another pivotal moment came in the form of portable electronics in the 1920s. The radio company Zenith introduced the first portable radio in 1924 ("Heritage"). However, models of this era were bogged down by heavy batteries and fragile glass vacuum tubes. This made them bulky enough to require a carrying case, such as the Zenith Trans-Oceanic model (Litwinovich). They were portable but not quite wearable.

It wasn't until the late 1940s and early 1950s that transistor circuitry allowed radios to shrink down into a handheld device. We can thank Sony's TR-63 in 1957 as the world's first pocket-sized radio ("Product & Technology Milestones"). Information that used to be confined to a wood cabinet in your living room or inside your car – news, weather, shows, and music – was now available on the go. It created an entire category of wearables exclusively for personal entertainment, spawning descendants in the form of the Walkman and later the Discman, MP3 players, and every music and podcast app.

Wearables as Always Connected

We can mark the rise of cell phones as the tipping point into constant contact. While the first cell phone in 1973 was the size of a brick, it was the beginning of truly mobile communication (McNamee and Yu). Once we cut the cord, there was no stopping our ability, much less desire, to always be in touch. Cell phones jumped into our cars, evolved into petit digital phones and flip phones for our pockets and purses, and had cousins in the form of pagers and BlackBerries. And then the smartphone arrived in 2007.

With honorable mention to the IBM Simon, Apple was the one that cemented the smartphone as a must-have device. They created a user experience that we hadn't seen before, paired with an ecosystem of devices seamlessly working together. This level of orchestration means that an iPhone is synced with the Apple Watch. It's possible to take a call from my wrist while my phone is in another room, all the while my AirPods are delivering the sound of someone's voice directly into my ears. We see this harmonization continue with the Apple Vision Pro, which accepts FaceTime calls before your eyes ("Apple Vision Pro").

The evolution of wearables is a majestic testament to human ingenuity. All of this progress is now making its way into our work lives. Particularly where safety is concerned, wearable technology has the power to shine a light on hidden risk.

WEARABLES AS PPE

Thanks to wearables, the work of safety is no longer the sole domain of the safety professional – it is now a crowdsourced effort. It's smart PPE, with the power to protect the individual user, everyone in their immediate area, and everyone in the larger workplace.

When we look at traditional PPE, they are exclusively about the individual. Without a doubt, we need static wearables like hard hats, hearing protection, steel-toed boots, and the likes. We want Jerry's head to be safe from falling debris and Jane's feet to be insulated against electricity. But no matter how many of these wearables are issued throughout a work site, they protect one person and only one person at a time.

The same is true of digital wearables, such as those that look for medical symptoms related to heat stress. They use metrics like heart rate, external body temperature, and sweat activity as signs someone is becoming overexerted. But as incredibly useful as these devices are, they continue to focus on the individual.

That's where our safety wearable takes a fundamentally different approach to risk – we look for hazards in the environment.

First, we collect environmental data that has a direct bearing on safety but is otherwise hard to detect. After all, you can't see with your own eyes when decibels are too high or air quality is poor. Our wearable intentionally doesn't have biometric readings or continuous location tracking because we know an accident is rarely the fault of one person or an isolated mistake. We're looking outward from the employee, not inward.

Second, our wearable forms a distributed network of environmental sensors. Each individual becomes a safety contributor, serving as the new "eyes and ears on the ground." Since readings are completely mobile, we can glean environmental insights far better than stationary devices. Every employee is continuously reporting on their environment as they move throughout the workplace performing their job. Companies can use this collective data to gain insights into ever-changing working conditions, allowing them to proactively avoid hazards and eliminate risk.

Wearable safety technology means that every worker is participating in an invaluable group project – one that could very well save a life.

KEY TAKEAWAYS

- Our bodies have worn technology for centuries, starting with the first glasses to today's smart watches
- All wearables create an exchange of information
- Safety wearables are a logical iteration of this technology

BIBLIOGRAPHY

"A Guide to History of Pocket Watches." *Watch Museum*, https://watchmuseum.org/a-guide-to-history-of-pocket-watches/

"Apple Vision Pro." *Apple*, 2024, https://www.apple.com/apple-vision-pro/

Beckman, Jeff. "15 Wearable Technology Statistics [2023 Edition]. *TechReport*, 30 May 2024, https://techreport.com/statistics/wearable-technology-statistics/#:~:text=Wearable%20Technology%20Sales,-5.&text=2022%20showed%20the%20number%20of,billion%20people%2C%20and%20sales%20increased

"1810, The First Wrist Watch." *Breguet*, n.d. https://www.breguet.com/en/history/inventions/first-wristwatch

"Heritage." *Zenith*, https://zenith.com/heritage/

Inspire, 2024, https://www.inspiresleep.com/

"Inventing the Transistor." *Computer History Museum*, 2024, https://www.computerhistory.org/revolution/digital-logic/12/273

"The Invention of the Cardiac Pacemaker." *U.S. Department of Veterans Affairs*, 2 August 2018, https://www.research.va.gov/research_in_action/The-invention-of-the-cardiac-pacemaker.cfm

Litwinovich, Paul. "Zenith Trans-Oceanic, The 'Royalty of Radios.'" *WHSU Public Radio*, 27 March 2015, https://www.wshu.org/vintage-radio/2015-03-27/zenith-trans-oceanic-the-royalty-of-radios

McNamee, Kai and Mallory Yu. "50 years ago, Martin Cooper made the first cellphone call." *NPR*, 3 April 2023, https://www.npr.org/2023/04/03/1167815751/50-years-ago-martin-cooper-made-the-first-cell-phone-call

"1952: Transistorized Consumer Products Appear." *Computer History Museum*, 2024, https://www.computerhistory.org/siliconengine/transistorized-consumer-products-appear/

"Product & Technology Milestones." *Sony*, 2024, https://www.sony.com/en/SonyInfo/CorporateInfo/History/sonyhistory-b.html

"75th anniversary." *Medtronic*, April 2024, https://www.medtronic.com/us-en/our-company/history.html

Van Welsenes, G. H. et al. "Improvements in 25 Years of Implantable Cardioverter Defibrillator Therapy." *Netherlands Heart Journal*, vol. 19, no. 1, 22 December 2010. *National Library of Medicine*, https://www.ncbi.nlm.nih.gov/pmc/articles/PMC3021193/

"Wilson Greatbatch: The Implantable Cardiac Pacemaker." *Massachusetts Institute of Technology*, https://lemelson.mit.edu/resources/wilson-greatbatch

6 Sophisticated Sensors

In our daily lives, we encounter technological marvels that not long ago seemed like mere fiction. Computing has become entirely personal as we now carry in our hands the same processing power that used to require rooms of equipment. This could never have happened prior to this point in history without the emergence of several key technologies.

Up until the past 20 years, collecting and analyzing sensor data at scale and in real time was difficult. We had to overcome a number of limitations with communication infrastructure, data streaming, device size, and power consumption. Now sensors are so small, accurate, and energy efficient that they can reside on the human body. Today's wearable safety sensors process large amounts of data for hours on end. Computing speeds and analytics then allow us to process data about what an employee is experiencing on the job and draw conclusions about working conditions as they're happening.

All of this wouldn't be possible without a number of inventions. It starts with the debut of the semiconductor, which led to integrated circuits. We also have binary code and data processing speeds to thank for the ability to draw insights. These milestones have all led to a powerful network of sensors backed by advanced computing power that is saving lives in the workplace.

SWITCHING TO TRANSISTORS

Today's wearables could never be achieved without the transistor. In 1947, a team at Bell Laboratories led by physicist William Shockley set out to unravel the mysteries of solid-state physics. At the heart of their exploration lay the dual nature of the semiconductor, which is part conductor, part insulator. Among the scientists, John Bardeen and Walter Brattain envisioned a world where these materials could act as tiny switches or amplifiers.

When they created the first functional transistor, it was a humble contraption: a few wires delicately attached to layers of gold foil and semiconductor material. Yet it could control the flow of electricity with unprecedented precision ("The Transistor"). This was a major evolution compared to the prevailing technology, which relied on bulky, power-hungry vacuum tubes (Zorpette). The transistor's ability to regulate currents directly resulted in the creation of electronics that are smaller, more efficient, and more versatile.

BIRTH OF BINARY

Transistors also gave rise to binary code, a language of ones and zeros that computers can process and store. Computers use transistors as switches to control these binary signals so they can be read. A single wire can signify one when it is carrying a current or zero without. Early computers, with their large heat-generating vacuum tubes,

DOI: 10.1201/9781003432302-8

were far less proficient at this two-symbol system. But with the advent of transistors, computers became more sophisticated while shrinking in size.

The rapid miniaturization eventually began to target everyday consumers, such as the first personal computers. Reducing the transistor's footprint was a crucial first step toward the proliferation of internet-connected devices all around us today. It's the reason why an industrial worker can don a safety wearable and seamlessly report data about their environment.

SAND TO CIRCUITS

When sand is processed, the refined silica creates the foundation for modern transistors, often referred to as silicon. Transistors can be laid down onto large sections of silicon to form dense groups, called integrated circuits or chips, working together to accomplish a task. Innovators have continued to find approaches for increasing the density of transistors on silicon for decades.

This rate of miniaturization was predicted by Gordon Moore in 1965 ("1965: 'Moore's Law'"). As the co-founder of Intel, he suggested miniaturization would happen at a rate of doubling transistor density every two years. This has largely held true. For example, the first Intel processor was released in 1971 and contained 2,300 transistors ("Intel's First"). In 1982, Intel released the 80286 processor, which touted 134,000 transistors ("Intel 45"). By comparison, the most recent release of Apple's M2 Ultra processor contains a staggering 134 billion transistors on a single silicon substrate ("Apple Introduces").

Semiconductor manufacturers have continued to develop innovative techniques to stack semiconductor materials to keep the miniaturization dream alive. A byproduct of Moore's Law is that the device housing can shrink as well. We no longer have computers that occupy a closet – they can be worn on our bodies as smart PPE instead.

PROCESSING AT LIGHTNING SPEED

In the early days of computing, machines were tasked with automating processes that were once painstakingly done by hand. As technology progressed, computers could perform increasingly complex and demanding tasks that would have been insurmountable for humans to perform manually in the same amount of time. Thanks to speeds like petaFLOPs, we have the computing bandwidth to collect and process vast amounts of data.

A petaFLOP is a unit of measure describing a quantity of computer calculations. An individual FLOP is a single floating point calculation. In simpler terms, each time you add two prices with dollars and cents together in your head, you perform a FLOP. The prefix "peta" is used to note a multiplier of a quadrillion (10^{15}). To put this into perspective, one petaFLOP is equivalent to doing a single math operation every second for over 31 million years ("Understand measures"). Currently, the fastest computer in the world is at Oak Ridge National Laboratory and can execute over 1,000 petaFLOPs per second; the second fastest comes in 585 petaFLOPS and is owned by the Argonne National Laboratory (Afifi-Sabet).

For safety professionals, these speeds are critical to accomplish the goal of understanding human work in real time. Along with miniaturization, they are what empowers wearable sensors to uncover indicators that are often hidden, like detecting indoor air pollution or risky movements that might lead to a repetitive motion injury or an unexpected fall.

VOC SENSORS FOR AIR QUALITY

Indoor air quality (IAQ) is one of the hardest safety concerns to detect as it is rarely visible to the eye or noticeable to the nose. Volatile organic compounds (VOCs) are chemicals that compromise IAQ when they evaporate. Often synthetic in origin, they are emitted by paints, cleaning agents, adhesives, sealants, and fuels. Even at low levels of exposure, health effects may include headache, dizziness, nausea, and fatigue as well as nasal, skin, eye, and throat irritation. VOCs are particularly insidious because they are often a cumulative hazard through repeat inhalation rather than a single exposure ("Volatile").

Thanks to the evolution of semiconductors, the latest generation of IAQ sensors is the size of a grain of rice. Our wearable uses a metal oxide sensor, which contains a semiconducting material that binds with oxygen. The presence of VOCs removes some of those oxygen molecules, decreasing resistance and increasing current flow. By measuring the resistance of the semiconductor, the wearable sensor can estimate the total VOC (TVOC) concentration in the environment.

Because these mobile IAQ sensors take readings wherever workers go, they are particularly adept at uncovering VOCs in surprising places. Case in point – our wearable helped a food production company find a hidden VOC source from a machine applying glue to cardboard lids. This was a shock as no one had reported fumes or health concerns. Furthermore, the glue's safety data sheet didn't have a VOC warning. However, when the company talked to the adhesive manufacturer, they learned that the glue doesn't off-gas once it cools. Without the IAQ sensor, no one would have realized that VOCs were present while the glue was warm.

ACCELEROMETERS FOR SAFE BODY MECHANICS

Accelerometers are based on Newton's first law of motion that an object will remain at rest unless acted upon by an external force. A type of accelerometer called microelectromechanical systems (MEMS) measures static forces such as gravity as well as dynamic forces resulting from motion or vibration.

MEMS accelerometers detect changes with a high level of accuracy. For instance, when an object experiences acceleration, it exerts a force on the accelerometer's sensing elements, causing them to shift with motion. This movement is measured and converted into an electrical signal, providing information about the object's acceleration. The applications are vast, from features such as screen rotation and stability control in helicopters to health applications such as tracking respiratory rate or fitness activity.

For workplace safety, accelerometers are how MākuSafe flags repetitive motion and slips. We use a three-axis accelerometer that correlates to typical motion: side to side, up and down, and back and forth. These movements are recorded over the course of a day. Using a calculated result called a signal magnitude vector, we can understand the cumulative impact of these three types of accelerations. That data is charted for individual workers in our Motion Explorer, which creates a profile that tracks and ranks motion exposure by hour, day, week, and month. This analysis is what allows a company to identify workers who have high potential for injuries.

For example, the data shows there's a difference in motion exposure for two people who are working in the same area and performing the same job. It's a clue that prompts safety leaders to investigate why there is a disparity. They can then implement meaningful changes, such as ergonomic training or adaptive equipment.

This data is a game changer. It isn't effective for a safety professional to conduct a walkthrough and be in the right place at the right time to catch a glimpse of someone moving in a risky manner. Instead, sensor data can highlight workers who have the greatest amount of risk. After making improvements, data can be collected again to show what impact was made. All thanks to a tiny MEMS accelerometer.

KEY TAKEAWAYS

- Key inventions like transistors and integrated circuits are the foundation for modern sensors
- Device miniaturization allows them to be worn on the body
- Computing speeds enable real-time data processing to uncover safety indicators

BIBLIOGRAPHY

Afifi-Sabet, Keumars. "The 7 most powerful supercomputers in the world right now." *Live Science*, 28 March 2024, https://www.livescience.com/technology/computing/top-7-most-powerful-supercomputers-in-the-world-right-now

"Apple Introduces M2 Ultra." *Apple*, 5 June 2023, https://www.apple.com/newsroom/2023/06/apple-introduces-m2-ultra/

"Intel's First Microprocessor." *Intel*, n.d. https://www.intel.com/content/www/us/en/history/museum-story-of-intel-4004.html#:~:text=Initial%20transistor%20count,technology%20held%20560%20million%20transistors

"Intel 45 Nanometer Technology Gives the Transistor Dramatic Make-Over." *Intel*, 29 January 2007, https://www.intel.com/pressroom/kits/45nm/IntelTransistor_Bkgndr_FINAL.pdf

"1965: 'Moore's Law' Predicts the Future of Integrated Circuits." *Computer History Museum*, 2024, https://www.computerhistory.org/siliconengine/moores-law-predicts-the-future-of-integrated-circuits/

"The Transistor: 75 Years Since the Famed Nokia Bell Labs Invention Changed the World." *Nokia Bell Labs*, 15 December 2022, https://www.bell-labs.com/institute/blog/the-transistor-75-years-since-the-famed-nokia-bell-labs-invention-changed-the-world/#gref

"Understand measures of supercomputer performance and storage system capacity." *Indiana University*, 2024, https://kb.iu.edu/d/apeq#:~:text=A%201%20petaFLOPS%20(PFLOPS)%20computer,every%20second%20for%2031%2C688%2C765%20years

"Volatile Organic Compounds' Impact on Indoor Air Quality." *EPA*, 13 August 2024, https://www.epa.gov/indoor-air-quality-iaq/volatile-organic-compounds-impact-indoor-air-quality

Zorpette, Glenn. "How the First Transistor Worked." *Spectrum IEEE*, 20 November 2022, https://spectrum.ieee.org/transistor-history

7 IoT for Safety

Currently, there are an estimated 17 billion internet-connected devices, which is over twice the global population (Vailshery). If we could visualize this distributed network on a map, it would appear that everything and everyone is connected. The information they generate is transforming our understanding of the world, including workplace safety in industrial settings. When safety sensors are worn by workers, the data can be extremely valuable in understanding how work is actually happening.

This result provides an unparalleled view into atmospheric conditions and ergonomic motion – leading indicators that contribute to the majority of accidents. By analyzing this pool of IoT data, safety professionals can uncover risks and hazards before they become a problem.

IoT RADICALLY CHANGES PROBLEM-SOLVING

One of the ways IoT is a game changer is that it opens up new ways of problem-solving. Thanks to our constant state of connectivity, we are increasingly asking, "why can't I know this continuously for myself and never miss an important piece of information?" Real-time data from IoT empowers us to gain those answers, using insights to tackle issues in innovative ways.

A prime example is weather data. There was an age when only scientists were in charge of collecting readings. They had to physically step outside, measure the relative humidity with a hygrometer, and then submit recordings to be aggregated into a forecast. Today, we have volunteer networks like Weather Underground that are based on IoT. Participants install their own weather station that includes atmospheric sensors such as wind speed, rainfall, pressure, and temperature. A WiFi device then transmits the data for analysis. With over 250,000 contributors, this crowdsourced IoT network continues to change the nature of weather forecasting ("Personal Weather").

We can also look at how people use health and fitness trackers to proactively make decisions based on health data. They don't have to wait for a yearly doctor's appointment to see that their wellness needs attention. They can voluntarily make adjustments to their diet, activity levels, and sleep hygiene to improve their numbers – all because a wearable smart device is providing actionable information.

Commercial and industrial buildings have pioneered IoT networks as well. Major systems such as lighting, HVAC, and energy mechanicals have embedded sensors. Because these devices are reporting on efficiency, maintenance, and performance, they're changing the way facilities are operated and even automated. Even the simplest solutions can be revolutionary. Imagine the labor savings of an IoT paper towel dispenser that can send an alert when it's low rather than a worker manually checking every bathroom on a schedule.

How does IoT support safety professionals? By leveraging workers to create a dynamic network that is always on the move. The downside of conventional sensors

DOI: 10.1201/9781003432302-9

is that they are typically mounted on a wall with a limited radius for detection. Even with handheld or tripod devices, a user may find it difficult to multitask while taking the reading. However, wearable safety devices will monitor for environmental and ergonomic hazards, specific to them, with no impact to workers' jobs. Their only involvement is to wear the device just like any other PPE. As a result, these wearable IoT sensors generate data about the immediate surroundings of your most precious assets – your people.

HOW IoT AUGMENTS SAFETY OBSERVATIONS

Having IoT sensors moving throughout your facility is an incredible advantage over walkthrough observations. Findings from site inspections are often limited because they only occur at specific intervals. Despite their vast training and knowledge, it's rare for a safety professional to be at the right place and the right time to witness a hazard.

Let's meet Josh, an industrial hygienist who is working hard to protect the front-line workers at his facility. There are several reasons why site observations aren't yielding the safety data he needs.

- **People change their behavior when they are being watched**. Workers know who Josh is and what his job entails. The minute he appears, they'll start being more careful. Except that Josh doesn't need people to be on their best behavior – he needs them to act normally. Just by his mere presence, Josh isn't getting an accurate view of everyday working conditions.
- **Josh has a preconceived idea about what risks could be present**. He's been to this facility many times and understands which hazards are inherently unique. He also knows its safety track record through dealing with past incidents. However, this familiarity can start to color his perception of what to expect, and he'll start having more difficulty looking for risk with fresh eyes.
- **Josh is just one person – he can't be everywhere at once**. What are the odds that he will catch something that is a safety issue? Even if Josh systematically performs his observation once a week for 30 minutes, that's just a fraction of time. The law of averages isn't working in his favor.
- **Workers can have a subjective view of safety**. For example, Josh could issue a survey asking about which day is the hardest for physical exertion. The majority of people might cite Monday simply because of its reputation, yet Josh sees that most injuries occur on Friday. Now he's confused because his workforce's perception about risk doesn't match his safety numbers.
- **Because the work environment is continuously changing, risk can be introduced at any moment**. But Josh can't be on the floor all the time – observations are only one part of his job. How can he track indicators when he's elsewhere?
- **Risk can be invisible to humans without instruments**. Josh can only rely on his five senses so much. He will inevitably need specialty tools to check for hidden hazards like decibel levels, VOCs, ambient light, and humidity. These readings all take time and training to perform, all of which are in short supply for Josh.

It leaves Josh wondering why he can't use technology to do his job in the same way as other professionals. He thinks about the warehouse operations manager who has vibration sensors on the conveyor systems. Or the day trader who works from her home office yet knows stock prices on a second-by-second basis.

Josh is thinking to himself, "How come I don't know at any given moment what's happening with the safety position of my company? Shouldn't I be able to know what kind of air quality my workers are experiencing in real time? Isn't it possible to collect data on fatigue contributors and then trend them for patterns? Where are the tools that allow me to see the bigger safety picture?"

Josh is asking the right questions, and he can gain these insights by equipping his workers with safety sensors. The resulting data supercharges his job, allowing him to oversee risk without physically having to be present.

With safety data generated at all hours and throughout the entire facility's work zones, Josh can now drill into the true nature of risk at this location. The details provided by environmental and motion sensors help him uncover several surprising issues.

Heat Stress

The Scenario

Josh's company has a challenge with indoor temperatures because it uses equipment that releases heat as a byproduct. The outside temperature is also impacting the indoor temperature despite their efforts to move air throughout the facility. This has resulted in a heat index that is regularly over 100-degrees Fahrenheit. It is such a pervasive problem that the company tries to stay ahead of dehydration by lining the hallways with stacks of sports drinks.

The Problem

Though the facility had three wall-mounted temperature sensors, they weren't IoT connected. Each thermometer has a clipboard hanging below, and it is somebody's job to walk over every hour and record the temperature. Occasionally, somebody takes those numbers to Josh, who tries to manually draw some conclusions. But ultimately, these readings aren't useful because they aren't automated or aggregated effectively. They also aren't representative of what each worker is experiencing.

The Response

Using wearable sensors like MākuSafe to gather real-time data directly from their employees, Josh has immediate insight into heat exposure and can anticipate when workers are going to need hydration breaks. Protecting workers moved from a manual, brittle process to one that is proactive, ultimately leading to a more productive and safer work environment.

Physical Exertion

The Scenario

Josh's company understands that the line between work as designed versus work as done is always moving. But they need a more intelligent way to measure these fluctuations. For example, they want to know if their risk profile changes throughout the day, such as the difference between eight in the morning versus two in the afternoon. This is driven by a rise in motion injuries among their loading teams.

The Problem

After a period of data collection, a trend emerges – every Thursday at 2pm, there's a significant spike in motion across the whole team. Because this window of time is a complete outlier, it's a concrete clue for Josh to investigate. In talking to the affected team, he learns that this concentration of movement matches the weekly delivery of a large shipment. Because it's a large load to move and the dock isn't air conditioned, everyone rushes to complete the work. Despite their efforts to be efficient, however, the team reports feeling hot and exhausted each time.

The Response

Josh not only knows that this one particular afternoon has higher risk, but more importantly, what is the root cause. The data pointed Josh to a trifecta of lagging indicators: heat exhaustion, dehydration, and physical motion injuries. He can now start to problem solve, such as working with the shipper on its schedule, adding more team members to this shift, or ordering automated conveyors for unloading.

Inadequate Lighting

The Scenario

Part of Josh's site includes a warehouse with forklift drivers. Even with regular safety training and OSHA certifications, there's still a considerable amount of accidents in this zone. Josh is stumped why this is an ongoing issue given that standard precautions are in place.

The Problem

After outfitting workers with wearable safety sensors, the likely root cause is revealed – this area is registering inadequate lighting. Upon visual inspection, Josh finds that due to recent volumes in orders, pallets of product are being staged in this area. The high stacks are significantly reducing visibility. It's also highly likely that workers are struggling to see fork trucks due to the lower lighting and tighter aisles.

The Response

Mitigating this risk started with creating a new staging area and relocating pallets. Josh also met with the forklift drivers to ensure they have a way to report issues like this in the future.

KEY TAKEAWAYS

- Workers with wearable sensors can be added to your IoT network
- Technology and data support the important work of safety professionals
- Data can uncover the true nature of risk

BIBLIOGRAPHY

"Personal Weather Station Network." *Weather Underground*, 2024, https://www.wunderground.com/pws/overview

Vailshery, Lionel Sujay. "Internet of Things (IoT) in the U.S." *Statista*, 4 June 2024, https://www.statista.com/topics/5236/internet-of-things-iot-in-the-us/#topicOverview

8 Machine Learning for Intelligent Insights

How do you handle the volume of data generated by safety sensors? Because IoT devices produce a prolific amount of information, traditional analytics aren't capable of analyzing its volume or complexity. We need artificial intelligence to shift through it and discover actionable information.

Machine learning has the ability to consolidate, organize, and analyze fast-moving safety data. It's what allows safety professionals to uncover important trends connected to risks and hazards. Using these intelligent insights, they can improve working conditions and have a profound impact on safety.

UNDERSTANDING INTELLIGENT INSIGHTS

The defining quality of an intelligent insight is context. Though many smart devices generate tremendous amounts of raw data, we need a framework to give it meaning. Otherwise, all you have is noise. An intelligent insight puts data to work.

- From data, we create information
- From information, we can draw insights
- From insights, we can make decisions
- From decisions, we can measure change

For example, a temperature reading is a type of raw data. By itself, that number doesn't mean anything. But an insight is created when we determine that the temperature is too high and poses a safety risk to workers. This intelligent insight empowers leadership to make adjustments that will reduce the likelihood of heat stress.

Intelligent insights are also a workforce multiplier because they help safety professionals find indicators where they may not have realized they existed in the first place. It's all about uncovering the true nature of risk, which is revealed by deviations in the data. What we've found many times is that what is typical or "normal" in a facility is just the status quo – no one has evaluated whether it's safe or unsafe.

Let's use noise as a safety concern. According to OSHA thresholds, 85 decibels or lower is permissible without additional safeguards ("Occupational Noise"). If your facility's average is above 85 decibels, a hearing conservation program is required. But sometimes companies haven't taken the time to establish their decibel average. They have no idea what their everyday levels are, much less where they fall within OSHA's criteria. Even if they start taking regular readings, they're awash in information that needs context and interpretation. For example, how would they know that

DOI: 10.1201/9781003432302-10

noise reaches harmful levels on Wednesday and Thursday afternoons? They need a way to corral all that data so they can derive intelligent insights from it. This is where they can look to artificial intelligence to help interpret data quickly and efficiently.

MACHINE LEARNING FOR SAFETY DATA

Machine learning, which is a subset of artificial intelligence, uses algorithms to train a model to recognize patterns and classify that data accordingly. It's how large amounts of data are processed so intelligent insights can be drawn in an automated way.

In *unsupervised* learning, we provide a plethora of examples from different data sets and allow the machine learning model to discover the nuances between them. This self discovery is often used for unexplored data sets to find previously unobserved groupings of data.

In *supervised* learning, we show the model examples of known data sets and then have it look for new examples that match. The MākuSafe team illustrates this at conferences with a simple potato and a touch of humor. We show a picture of three bins of potatoes labeled Group A, B, and C to match their colors. Then we show a picture of an unlabeled potato and ask the audience which group it belongs to. When they classify it correctly, we joke that they just performed machine learning. This is because each audience member "trained" themselves by looking at the picture of potatoes in bins. In fact, this is supervised learning because similar data was already grouped into the bins. When presented with an uncategorized potato, they could immediately classify the potato based on the similar ones from the picture.

Machine learning models are trained using thousands of examples that are grouped into known classes. For example, models have been populated with millions of handwritten letters so they have the ability to decipher individual letters. When the model is presented with a new handwritten sample, it knows an "a" from "o" and so on. More importantly, this can be done at scale across millions of examples with amazing speed – no human intervention is required.

This same technique is changing safety. It's essentially how MākuSafe has trained our motion model to recognize slips, trips, and falls. When somebody has a high force motion, our machine learning model can say "I recognize this motion signature as a slip." It is even able to tell us how confident it is with the classification: "I'm 73% confident that is a slip, 17% confident it's a trip, and 10% confident it's a fall." Not only is this useful for uncovering individual instances, but safety leaders might also discover a cluster of slip incidents that signal there is an underlying problem at play.

TRENDS UNCOVER ISSUES

Part of the beauty of machine learning is that it can discover trends that may not be perceived easily by safety professionals or uncovered by traditional data analytics. This is because it's difficult to interpret a single point on an IoT network, much less all of them combined. So, we use a consensus engine to shift through sensor information, tabulating their signals into a voting system that creates meaning.

Air quality is a common example of a hidden issue. How can you truly capture if an area has high VOCs? It can be difficult to discern if there's a genuine problem or an outlier, like a worker who put on too much body spray. By using data to establish a trend, it's easier to show when a pattern has grown into a concern worth exploring.

Let's say there's a trend that shows air quality has decreased by 26% on your loading docks. We can dig into that, noting that CO_2 went up 14.2% over the last three weeks while VOCs went up 25.9%. We can next look at the time frame to see how long this issue has been occurring. We can see that in the last three weeks, it's risen to 600 ppb. Knowing that our baseline is 510 parts per billion, that's a 15% rise in VOCs.

All of this context from the data is what leads us to an intelligent insight – there is proof there is an air quality issue and we need to investigate. A logical step would be to look into the ventilation system or see if something unexpected is offgassing. For example, the root cause is that the dock floors had just been painted. By adding some temporary fans to help dispel fumes, VOCs will soon return to normal.

GETTING AHEAD OF TRIPPING HAZARDS

Slips and trips are notoriously difficult to document, mainly because most people brush it off as inconsequential. They might even assume they are at fault – who hasn't sheepishly tripped because they were shuffling their feet? But what that one individual doesn't realize is that eight of their coworkers have also tripped at this particular spot. It's possible the culprit is something in the environment, like an uneven floor, that needs attention. How do you gain optics into this risk and have the ability to minimize it before it escalates into a fall and someone gets a broken wrist or worse?

Here's how it might look in practice. The data from our wearable device notes an increase in slips in a particular concrete section. An analysis suggests that it may be the result of increased condensation from humidity at a certain time of day. This intelligent insight allows the health and safety manager to collaborate with the facilities team to run the air handler more frequently in that area.

One client experienced this capability firsthand and mitigated a hazard that could have caused a slip or trip. MākuSafe data captured several slipping motions at one entrance. When the safety team went to investigate, they found that snow had blown onto the sidewalk leading up to the building. They worked with their site logistics team to clear off the path and apply ice melt.

This is the power of using advanced analytics and machine learning to handle safety data. Far from replacing safety professionals, a machine learning model allows them to rapidly digest and process data. Armed with intelligent insights, leaders are empowered to make safety decisions in new and profound ways.

KEY TAKEAWAYS

- Data needs context in order to be actionable
- Machine learning models help organize and process data
- Trends in the data can reveal safety issues that are otherwise hard to detect

BIBLIOGRAPHY

"Occupational Noise Exposure." *Occupational Safety and Health Administration*, https://www.osha.gov/noise#:~:text=OSHA%20requires%20employers%20to%20implement,%2Dweighted%20average%20(TWA)

Part III

Leadership, People, and Implementation, Tom West

9 Leading Indicators from Wearables Improve Safety Outcomes

In November of 2019, we had the opportunity to participate in a Public Stakeholder Meeting held by the Occupational Safety and Health Administration (OSHA) on "Using Leading Indicators to Improve Safety and Health Outcomes." The meeting was a follow up to a June 2019 OSHA publication that provides guidance and information to employers on incorporating leading indicators into their safety and health programs. Comments made throughout the session were supportive of the use of leading indicators in safety and health programs. Many trade organizations and larger companies had begun implementing or assessing the use of leading indicators, but there was a high degree of variability. Although there was no real standard for use at that time, and none still now, there is great support for OSHA and other entities globally to further promote their use to improve workplace safety outcomes.

A takeaway from this meeting was that there was agreement in spirit on the importance of leading indicators, but an acknowledgment that industries are different and businesses are unique. Therefore, interpretation may vary, and deciding which leading indicators were most meaningful to each would be challenging. Further, there seemed to be an absence of clear ways to use leading indicators, even when well defined.

Comments included what general attributes "good" leading indicators should possess. Frankly, it was disappointing that the suggested indicators were limited to things like training time, toolbox talks, daily safety conversations, meeting attendance, time spent in plant by senior managers, unsafe acts, scorecards, and the like. While tracking these is undoubtedly a step in the right direction, wearables and sensor monitoring are accessible now and offer a tremendous opportunity for real-time insights on experiences and exposures associated with leading indicators.

DEFINING LEADING AND LAGGING INDICATORS

One particularly insightful written contribution to that OSHA meeting came from the partners of ORC HSE Strategies, global thought leaders in health and safety. Their comments, on behalf of their significant roster of member companies, underscored the need for practical, actionable leading indicators that truly enhance workplace safety. Their perspective was spot on, highlighting the necessity for leading indicators to not be just metrics, but also "tools that drive meaningful safety improvements." They suggested tracking observations, safety suggestions and implementation rate, safety action items completed on time, building transparent cultures of finding and

DOI: 10.1201/9781003432302-12

fixing issues, focus on where the employee meets the work, risk-based approach, taking action and proactivity.

MākuSafe also voiced during that OSHA meeting that technology, specifically wearables, have the potential to generate tremendous amounts of new data that will offer employers easy access to meaningful leading indicators of risk and hazards. So much more is possible now, so easily, and provides significantly more potential to make an immediate difference.

Keep in mind that leading indicators are proactive, preventive, and predictive measures that provide information about the performance of safety and health activities. They focus on what could potentially cause future incidents or injuries. Examples include safety training sessions and safety audits, but more importantly the reporting of near-misses and anything that we can do to identify potential hazards before incidents occur.

Lagging indicators, on the other hand, are reactive measures that provide data on incidents that have already occurred. These include statistics on injury rates, lost workdays, and workers' compensation claims. While lagging indicators are crucial for understanding past performance, they do not predict future incidents. The idea of a balanced scorecard of both types is often suggested as it makes sense that lagging indicators should provide a useful gauge of how well our leading indicators have been selected and are working.

Leading indicators are essential for creating a safer workplace as they help organizations identify and mitigate risks *before* incidents occur. They contribute to a proactive safety culture and can lead to significant improvements in safety performance, such as:

1. **Early Detection of Hazards:** Leading indicators can reveal potential hazards early, allowing for timely interventions.
2. **Improved Safety Culture:** Focusing on leading indicators can promote a culture of continuous improvement and proactive safety management.
3. **Enhanced Employee Engagement:** Employees are more likely to engage in safety programs when they see their efforts in reporting and mitigating hazards are valued and acted upon.
4. **Reduction in Incidents and Injuries:** By addressing potential risks early, organizations can reduce the number and severity of workplace incidents and injuries.

Despite the clear benefits, implementing leading indicators is not without challenges. For starters, defining universal indicators is difficult since different industries and operations have unique risks. Next, management's buy-in to support and act on leading indicator data is crucial for success. The ability to collect and analyze leading indicator data can also be resource intensive and may require new technology or processes. Finally, successful implementation requires a positive safety culture where employees feel comfortable reporting potential hazards without fear of retribution.

OSHA and others have developed guides and materials with recommendations on using leading indicators. However, they acknowledge that, given the diversity of industries and operations, one-size-fits-all practices would be challenging to establish. Further, the vast majority of examples of leading indicators still being discussed in most organizations today seem to revolve around things that may be easy to measure but are not the most beneficial. Nearly every discussion of leading indicators seems to include things like measuring attendance at safety meetings, documenting training completion, and number of toolbox talks delivered. While these measures are important, they often fail to provide the most actionable and important insights needed to prevent incidents before they occur.

Given the work we were doing at MākuSafe and how we could contribute to meaningful leading indicator data, we developed our own acronym for characteristics of useful leading indicators – T.R.U.E^2. To be as useful as possible, leading indicators can and should have these characteristics:

Timely: The data should be available in real-time or near real-time.
Relevant: Indicators should relate directly to worker health and safety, including the most common causes of incidents, injuries, and claims.
Useful: The data should not only provide unique insights not easily obtained from other sources but also be immediately actionable.
Easy and Economical: It should be straightforward to gather and interpret the data without being prohibitively expensive.

The data gathered by the MākuSafe wearable devices includes an array of potential environmental hazards, human movement, and ergonomic risk and facilitates communication of frontline worker observations. Other wearables may only offer a portion of these, have different strengths, or specific applications. But in our case, the data represents leading indicators for a wide array of the most common and frequent workplace incidents and injuries. Additionally, by gathering both ergonomic and industrial hygiene data simultaneously, you can begin to understand if factors in the environment may be directly contributing to ergonomic risks. For example, poor air quality may be accelerating fatigue in workers, leading to a higher likelihood of MSD injuries from those exposed to the poor air quality.

T.R.U.E.^{2} IN PRACTICE

Consider this:

> "In the U.S. alone workplace injuries cost businesses more than $1 billion per week. That's more than $58 billion per year, with the vast majority caused by just 10 common injuries. Liberty Mutual Insurance regularly publishes their Workplace Safety Index (WSI) where they examine the top 10 causes of the most serious disabling workplace injuries – those leading to more than five missed workdays – ranked by their direct cost to employers based on medical expenses and lost wages. The top 10 are responsible for 82% of that $58 billion"
>
> (2023 Workplace).

FIGURE 9.1 Wearable technology can gather leading indicator data on a wide array of common but costly EHS concerns. This can include environmental hazards, such as temperature, heat index, noise exposure, and air quality. In addition, potentially harmful human movement and motion data can be gathered, including forceful strain and exertion, push/pull, and level of physical effort expended to perform work, all of which can contribute to cumulative physical impact and potential for MSDs.

Based on BLS data, the WSI identifies a number of interesting findings. In nearly every instance, the risk of these accidents could very likely have been mitigated by a leading indicator:

- Three of the top ten causes are related to slips, trips, and falls, which constitute 35% of the cost of the top 10 or about a third of the $58 billion total. How many of those were preceded by others who slipped, tripped, or fell in the same place without being hurt and didn't tell anyone? In the case of falls from height, how many of those had movement or motion indicators detectable by sensors from when others caught themselves on a wobbly ladder, or similar, but didn't fall?
- Two more causes out of the ten involve exertion, human movement, or ergonomics and are responsible for approximately another 35% of the cost of the top 10. Sensors like MākuSafe can identify forceful exertion and generate notifications for those workers who rank high in work effort being expended compared to their peers in similar work roles. Some wearables require a multitude of sensors and may be less practical for day-to-day wear, but can help identify sub-optimal postures or bends. In either case, these precursors can provide direction on where an ergonomic assessment is needed or even where video-driven AI can be deployed to further understand and prevent ergonomic injuries.

- Two causes – struck by or caught in – account for 7%. How many of those were near-misses where forceful or accelerated human movement, such as someone jumping out of the way or pulling their hand out, could have been detected before these injuries occurred?
- Pedestrian vehicular incidents also accounted for $1.61 billion. Since these are workplace injuries, it's fair to assume a good portion of those involve fork trucks or similar powered industrial vehicles. How many of those had traffic patterns, near-misses, or accelerated movement indicators that could have been identified by sensors before injuries result?
- One cause responsible for exposure to harmful substances costs $3.35B per year. How many of those could have been detected by VOC or gas monitors in advance of exposure?

A wearable device such as Mākusafe has been developed, which is able to contribute leading indicator data automatically when near-misses occur. This data, made useful, has predictive value that when acted upon can contribute significantly to all of these accident causes.

KEY TAKEAWAYS

- The 2019 OSHA meeting was important for MākuSafe because it reinforced the importance of leading indicators in improving workplace health and safety outcomes.
- It also validated our early thinking on leveraging real-time data to predict and prevent incidents.
- A focus on the characteristics of meaningful leading indicators enabled through sensors and wearables will be crucial in fostering safer and more proactive workplace environments.

BIBLIOGRAPHY

"2023 Workplace Safety Index." *Liberty Mutual*, 10 Aug 2023, https://business.libertymutual.com/insights/2023-workplace-safety-index/

"Comment from Newell, Stephen, et al; ORCHSE Strategies, LLC (ORC HSE)." *Regulations.gov*, 6 Feb 2020, https://www.regulations.gov/comment/OSHA-2019-0005-0023

Occupational Safety and Health Administration. "Meeting Summary: Using Leading Indicators to Improve Safety and Health Outcomes Public Stakeholder Meeting." *Regulations.gov*, 13 Jan 2020, https://www.regulations.gov/document/OSHA-2019-0005-0020

"Using Leading Indicators to Improve Safety and Health Outcomes." *OSHA*, https://www.osha.gov/leading-indicators/research

"Using Leading Indicators to Improve Safety and Health Outcomes – Additional Resources." *OSHA*, https://www.osha.gov/leading-indicators/research

10 Successfully Implementing Innovative Technology in Organizations

Introducing innovative technologies like wearable safety devices into an organization can be a transformative journey. Any digital transformation initiative, including safety technology, is broad and complex, requiring strategic planning and effective change management from leaders.

Organizations will also need to recognize that they may encounter employee resistance when implementing new technology. It's common for concerns like privacy or general apprehension to arise, which is echoed in the Safety Technology 2024 survey by The National Safety Council's Work to Zero initiative ("Safety Technology 2024"). Workers may not even see the value: "46% reported they did not want to change the safety equipment they were currently using and 41% reported current safety equipment and technology meets their needs so there is no need for new adoption" ("Safety Technology 2024"). And yet the report calls out that the biggest motivator for exploring new safety technology is a recent serious injury or fatality.

The foundation of implementing any new technology begins with a willingness to embrace change and a piloting attitude. Organizations must be open to exploring new ideas and experimenting with novel solutions so innovation can thrive. This chapter will explore the essential steps and considerations for successfully implementing such cutting-edge solutions like wearables. We'll delve into the importance of a piloting attitude, ensuring organizational readiness and navigating change management.

HOW TO ESTABLISH A STRATEGIC PILOT

Pilots serve as valuable testing grounds to assess the feasibility and effectiveness of new technologies before committing to full-scale implementation. In the early stages, it can be tempting to say yes to any pilot opportunity. However, it is crucial to recognize that managing successful pilots demands significant resources. Therefore, focus on well-defined pilots where both the vendor and the client can succeed. The pilot should have clear success criteria and the potential for scalability. As Stephen Covey taught us, "begin with the end in mind," or at least have partial clarity around problems to be addressed, and then define what a new solution can and should solve. Also, embrace the realities of the solution being considered, in its current format,

DOI: 10.1201/9781003432302-13

as well as the potential future capabilities that will be useful once developed and released by the vendor.

Both the vendor and the client need to prove the value of the new technology swiftly and efficiently. There is a phrase often used in startup circles called pilot purgatory. If a piloting client and the vendor haven't clearly identified how to proceed with agreed upon milestones, some pilots can go on indefinitely. It is in the best interests of both parties to understand the plan's initial stages, create measurable goals, and set a defined timeline. They should also agree on a long-term vision of what a larger relationship could look like. It's crucial to answer questions like:

- How will we measure success during the pilot?
- What is our time frame?
- Who needs to be involved *DURING* the pilot that would be responsible for scaling the technology if the pilot is deemed successful?

Choosing the right pilot site or department is crucial for a successful implementation. It is essential to identify areas with a high likelihood of success and minimal compliance hurdles. For example, we started one pilot with a food packaging operation in the warehouse with material handling workers. This allowed them to evaluate a trial run first, then scale up to cover more of their workforce in other parts of the operation.

Having an internal champion will also significantly impact the success of a pilot. Often holding titles like VP of Safety or Director of EHS, these advocates promote the new technology within the organization, staking their reputation on its potential benefits. They play a crucial role in driving change, rallying support, and ensuring that the technology is embraced by all stakeholders. Their commitment and dedication can make a substantial difference in overcoming resistance and achieving successful implementation. This support needs to trickle-down to frontline leadership, who will play a crucial role in making sure that the technology is being utilized and that the workers understand their role in the pilot's success. We'll expand on this in a moment, but it bears repeating that the role of the frontline leadership and frontline workers is extremely critical in properly evaluating the new technology.

Lastly, effective implementation requires cross-functional communication and early involvement of all relevant departments. IT, legal, HR, and other stakeholders may need to be engaged from the beginning to address concerns and ensure smooth integration. There are many people within an organization who need the opportunity to provide input or ask questions. A lack of stakeholder involvement can lead to significant delays and potential roadblocks. Open communication and collaboration will allow organizations to navigate these challenges effectively.

ENGAGING FRONTLINE WORKERS

Frontline workers are often the most affected by new technology implementations. Ensuring their inclusion and early communication, especially around why the change is important, is vital to avoid resistance and negative reactions. Listening to their concerns, addressing their questions, and involving them in the process develop a sense of ownership and acceptance. Focus groups and learning sessions can be valuable tools to gather feedback and demonstrate the organization's commitment to a smooth transition.

This is essential to ensuring both initial and ongoing usage or participation. As the saying goes, even the best tool in the world does no good if it sits in the toolbox unused. For technologies like exoskeletons, it may be feasible to obtain only a couple of units and ask for volunteers who want to experiment with them on their own for a few weeks, and have them report back.

In our case, to expedite time to value, we recommend deploying 100 wearables to a group of 200 workers on two different shifts, with client leadership committing to at least 80% participation. In about 100 days, we will have monitored approximately 100,000 to 150,000 hours of work, gathered millions of data points showing what normal work looks like as well as indicators of potential hazards and risk, and typically find loss avoidance evidence.

With good involvement and communication, this can be done positively. When pilots are enforced through mandatory compliance, they can be met with resistance. Clearly communicating the functionalities, limitations, and direct benefits to frontline workers can significantly enhance participation, ultimately ensuring the pilot's success. Leadership's involvement in this process is crucial for building trust and commitment. Early communication campaigns can also play a significant role in a successful launch. When these requirements are addressed, use of a new technology solution can be expected by leadership which leads to faster time-to-value than making use optional.

FIGURE 10.1 Installation of wearable kiosks in a large facility with hundreds of workers. To overcome the challenge of nearly 20 languages being spoken on site, card readers were installed on each kiosk to provide a swipe-and-go experience for workers.

Our solution requires collaboration with the client to preconfigure hardware, set up software instances, and then go onsite to do an installation, which usually takes about a half a day even for a large facility with millions of square feet of coverage area. We have learned that deploying our technology should coincide with an in-person overview directly. We underscore what the technology does, what it does not do, what data is gathered, and how it will be used. This is most successful when leadership is shoulder to shoulder with us, committing to the same tenants. In some instances, management even delivers this orientation themselves, preceding it with lots of early communication that new tech is coming soon, why it is important, and how it will be used. Some clients have built early communication messaging campaigns and were extraordinarily successful at launch as a result.

As the pilot progresses, ongoing communication between leadership and frontline workers is essential to maintain momentum and celebrate progress. Sharing data, demonstrating its use, and reinforcing the importance of the technology help in sustaining enthusiasm and commitment. Celebrating small successes and milestones creates a positive feedback loop that encourages continued adoption and utilization.

NAVIGATING UNIONIZED ENVIRONMENTS

Implementing new technology in unionized environments requires careful navigation. Engaging with both labor leaders and management early in the process can help address concerns and foster collaboration. By highlighting the mutual benefits of the technology, organizations can gain support from both sides and create a conducive environment for successful implementation. One illustrative story from our experience stands out.

We were in talks with a multinational heavy industrial manufacturer with a nearby unionized site about a potential pilot location. They requested an onsite visit and a joint meeting with labor leaders and members of management, which was a great idea as it swiftly led to bridging any gap between the two perspectives. After just a few minutes of discussion, something remarkable happened. Both sides of the table – union representatives and management – began volleying back and forth about who cared more for the workers on the front lines and who believed more strongly in the benefits of our technology.

This experience underscored the importance of clear, transparent communication and collaboration. By addressing the concerns and interests of both labor and management, we were able to help both sides see the value of the new technology from their unique perspectives.

KEY TAKEAWAYS

- Intentional and strategic change management is necessary to successfully implementing innovative technology like Mākusafe's wearable safety devices.
- Pilots should be designed with specific objectives, timelines, and success metrics.

- Gaining the support of workers, including union representatives, is critical when introducing any new technology.

BIBLIOGRAPHY

"Safety Technology 2024: Examining Trends in Technology Solutions Used to Reduce Serious Injuries and Fatalities in the Workplace." *National Safety Council*. n.d. https://www.nsc.org/faforms/work-to-zero-safety-technology-2024

11 Moving Beyond Behavior-Based Safety Approaches

In our conversations with various organizations, it's clear that many remain committed to Behavior-Based Safety (BBS) programs. That's understandable, as the journey of implementing successful programs is lengthy, involving deep organizational changes, and the intent and objectives behind BBS are undeniably worthwhile. However, while the theory of BBS may seem sound, its practical application often falls short of expectations. This chapter delves into the origins and principles of BBS, its potential benefits, and the common pitfalls in its implementation. It also explores why gathering input from frontline workers as a form of a leading safety indicator is invaluable, why that may be the most desirable behavior change, and how wearable tech supports this.

THE RISE OF BEHAVIOR-BASED SAFETY

The origins of behavior-based safety (BBS) can be traced back to the 1930s with the research of Herbert William Heinrich, an assistant superintendent at Travelers Insurance Company. Heinrich's research suggested that around 90% of accidents,

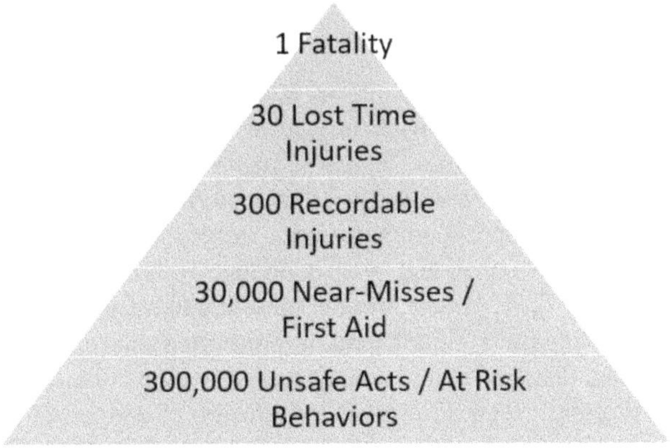

FIGURE 11.1 This illustration of Heinrich's Safety Triangle represents the concept that for every one fatality, there are typically 29–30 minor injuries, 300 recordables, 30,000 no-injury incidents, and 300,000 unsafe acts or at-risk behaviors. While this hierarchy emphasized the importance of addressing even minor incidents to prevent more serious injuries, it also unfortunately led to targeting worker behavior.

DOI: 10.1201/9781003432302-14

injuries, and illnesses were the result of "worker errors." This approach to health and safety, which assumes that most injuries and illnesses are caused by unsafe worker actions, became popular in the United States. Many safety professionals consider Heinrich a pioneer of modern BBS programs.

The term "BBS" was coined in the 1970s by Gene Earnest and Jim Palmer, two Procter & Gamble safety professionals working on a project to reduce injuries at their company. BBS is a proactive safety approach that builds on American engineer William Edwards Deming's principles of continuous improvement. It's also based on scientific research and writings by B.F. Skinner, the father of behavior analysis.

The premise of BBS is simple: By reinforcing positive behaviors and addressing unsafe actions, organizations can reduce workplace incidents. Early adopters saw significant improvements, leading to the widespread adoption of BBS in various industries. The ideal principles of BBS have good intentions:

BBS Actions	BBS Outcomes
Observation and Feedback: Regular observation of employee behavior, followed by constructive feedback.	**Reduction in Accidents:** By encouraging positive behaviors, many organizations have seen a decrease in workplace accidents.
Positive Reinforcement: Encourage safe behaviors through rewards and recognition.	**Improved Safety Culture:** Engaging employees in safety processes.
Employee Involvement: Engage employees in safety processes to foster ownership and accountability.	**Enhanced Communication:** Routine observations and feedback improve communication between employees and management.
Data-Driven Decisions: Use data from observations to identify trends and areas for improvement.	**Continuous Improvement:** Ongoing efforts to enhance safety protocols based on feedback and observations.

THE PROS AND CONS OF BBS

Despite the potential benefits, the reality of BBS implementation is often less than ideal. Instead of fostering a proactive safety culture, BBS can devolve into a bureaucratic exercise. One common issue is the reduction of BBS to a mere checklist. Employees are required to document a certain number of behavioral observations weekly, which can lead to superficial compliance rather than meaningful engagement.

Consider this scenario – you and your colleague Bob are required to document five observations each week. Under pressure and due to a friendly relationship, you both might end up filling out an index card or a form in a smartphone app about each other without diligently performing genuine observations. This act is sometimes called "pencil whipping" because it gives the illusion of a sign-off but there's been no real verification. The result is that Bob documents nothing but exemplary behavior from you, and of course, you return the favor for Bob. This undermines the purpose of BBS as the data collected becomes meaningless.

Another major pitfall is the presence of a blame culture. If employees fear discipline or blame for reporting unsafe behaviors, the entire BBS process becomes ineffective. The goal of BBS is to celebrate positive behaviors and identify improvement opportunities, not to single out individuals for punishment. When fear is present, employees are less likely to report honestly, rendering the system useless.

MOVING BEYOND BBS

While many organizations struggle with BBS, some achieve tremendous results by implementing it correctly. These organizations often engage all levels of employees in the feedback process without the threat of punishment. They'll also promote positive behavior by focusing on catching employees doing the right thing, or even approximately right, akin to the principles outlined long ago by Ken Blanchard in the book *One Minute Manager*.

Perhaps the most crucial behavior that we want and need from our people is open and transparent communication. That starts with creating an environment where employees feel safe to report issues and suggest improvements. Just imagine how your organization would benefit from truly unfiltered knowledge. Imagine what you could routinely learn from the frontlines by asking genuine questions like:

- What's on people's minds?
- What's going well and what needs to be changed?
- When reporting potential problems, what do they worry about?
- What tools or equipment do they need?
- If you had unlimited resources what improvements would you make in your job or your department, and why?
- How should the process be changed?
- What will go wrong next?
- Where do they think the next incident will occur?
- Where might someone die on the job?
- Where do we say "things are done this way" but actually do them another way?
- Where are we forced to do things outside of our plan because of constraints or other unplanned changes?
- Do you feel safe doing this job?
- If you could make this task easier, how would you do it?

Employees should feel comfortable sharing their thoughts, reporting potential problems, and suggesting changes without fear of reprisal. This kind of communication is essential for identifying and addressing risks proactively. Yet leaders continue to struggle to value and embrace this behavior.

We've asked live audiences for years now – how many believe that communication from the workers perspective is the "holy grail" of keeping people safe? Most or all hands in the room go up. Next, with hands remaining up, we ask how many have implemented a simple and structured way to collect this input – most hands go down. The next question is how many of you get as many reports as you would like or learn about as many things as you think actually happen? Now we're down to one or two hands at best. When asked how many reports they collect, the answer is usually around 8–10 while the highest is 20 a month or so. This demonstration leaves the rest of the safety professionals in the room in awe of those who are yielding that many.

FIGURE 11.2 Example of a construction worker using wearable tech to report observations and safety concerns via voice from the field.

This is how we met Mark, who is a remarkably experienced and excellent safety director in a large packaging company. Mark was the sole person with the last hand raised. He said his company received nearly 1,000 frontline reports a month from over 60 sites. They were distributed to senior management who actively responded, creating a culture of open communication and continuous improvement. He even credited this practice as being instrumental in helping them operate their business through the pandemic when no one was traveling.

Mark explained that he had a high degree of interest in wearable technology, especially if it could help facilitate communication from the frontline worker. Making reporting as easy as a push of a button could have the potential to increase the feedback they were already getting. Furthermore, if the software technology associated with the device could also help facilitate where those messages were directed and even automate actions for people in his organization, they could exponentially leverage the communication from the frontline workers. Mark knew that communication was crucial to building their safety culture yet he also knew he could always use more.

THE ROLE OF TECH IN FRONTLINE COMMUNICATION

MākuSafe's human-centric approach to wearable tech allows workers to push a button and report issues verbally through a device on their arm. Even though there's a wide array of sensors monitoring a plethora of experiences and exposures in real time, the MyVoice feature may be one of the most beneficial capabilities the solution offers. This innovation not only makes it easier for employees to report near-misses and observations but also ensures that these reports are captured accurately and managed efficiently.

MyVoice allows frontline workers to send real-time voice reports directly to leadership that is immediately actionable. This push-to-talk capability leverages the natural human ability to sense and report work conditions, making it a critical tool for workplace safety. Workers can report anything they see, smell, or feel that could be a hazard, providing valuable insights that wearable sensors alone cannot capture. This includes everything from near-misses, good catches, and physical or behavioral observations to maintenance requests and tool or workstation needs.

For instance, one of the first MyVoice reports received was a report about a missing guard on a machine. Rather than needing to stop what they were doing, hunt down a supervisor, or walk to another building to fill out a form, they only needed to take a few seconds to communicate their observation. And because leadership had this frontline observation, they could take action and likely prevented a serious injury. Another early message was the voice of a worker that was clearly shaken after she reported that she was nearly "struck by one of the machines" in their facility.

Our technology ecosystem prioritizes these messages above all other data collected by the wearable device. This approach underscores the importance of human observation in maintaining workplace safety. Empowering frontline workers to share their observations reinforces their role as the first line of defense in identifying and mitigating hazards.

Every company will say that they want to hear more and know more from the front lines. They try note cards, open office policies, suggestion boxes, and so many other ways to get clear and immediate information. A sophisticated organization knows that their frontline workers know more than anyone else in the organization about how work is being done. The burden often falls on the worker to find a supervisor, fill out a form, or stop work and report some sort of situation. Company productivity goals are often at odds with these solutions because they require a lot of that

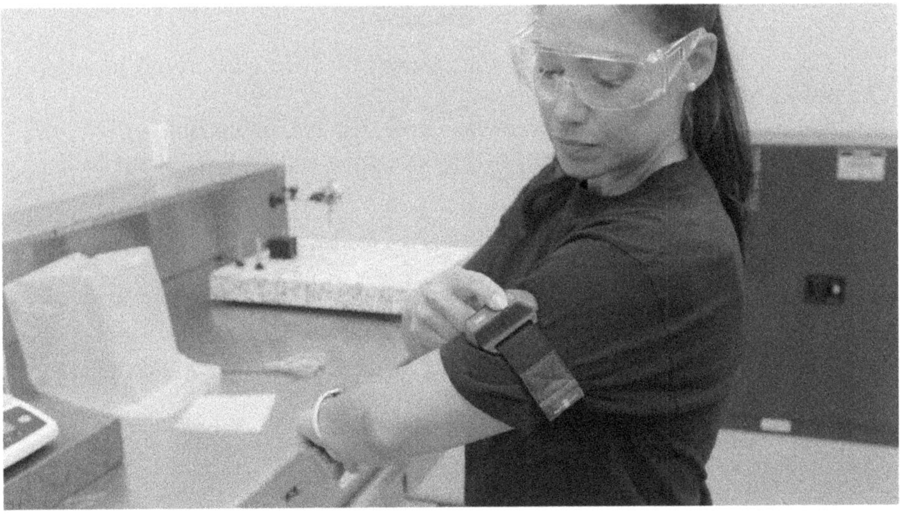

FIGURE 11.3 A worker can communicate observations to leadership by speaking into a wearable device.

worker, including sacrificing productivity. What it boils down to is leadership taking swift action on the observations of a frontline worker. It benefits the worker, the productivity, and ultimately the company. And every worker appreciates when their voice is heard.

It's worth considering also that if used well, these messages inherently possess psychological safety. The worker can volunteer to share what they see without worrying about any repercussions. This will be reinforced when leadership responds to their concern by redesigning work or mitigating a hazard rather than focusing on a person's behavior.

Finally, voice communication provides an incredible opportunity for a leader to respond by saying, "thank you, I heard you, and here's what we're doing about it." Companies should always take the opportunity to positively reinforce the fact that someone took time to communicate something they thought was valuable. We need more of that.

BUILDING A CULTURE OF OPEN COMMUNICATION

To evolve workplace safety, we need to move beyond BBS. The most important behavior to encourage in workers is a willingness to share their perspective. By fostering open communication, leveraging human-centric technology, and focusing on positive reinforcement, organizations can create a proactive safety culture that truly values employee input and continuously strives for improvement. The use of voice communication will further empower frontline workers, prioritizing risk identification and reporting as foundational elements of a successful safety program.

KEY TAKEAWAYS

- Avoid common pitfalls such as the checklist mentality and a culture of blame
- Create an open environment where employees are encouraged to report safety observations
- Focusing on work-related problems rather than individual faults
- Collect frontline observations by implementing real-time voice reporting

12 Haptic Feedback Isn't Always the Answer

The world is full of haptic technology. We're accustomed to devices that vibrate for alerts and touchscreens that click like a button. These sensations are found in everything from consumer goods like phones and watches to the entertainment world with video games and virtual reality. They're helping doctors learn to perform robotic-assisted surgeries and pilots to simulate flights. So wouldn't a mild pulse as a safety warning be just as helpful?

The surprising answer is "it depends." Haptic feedback can be useful for short-term compliance, such as in a life-threatening situation like an alert when dangerous levels of a toxic gas are present, or to evacuate immediately. It can also be helpful when a powered industrial vehicle or forklift is around the corner, but the view is obstructed. But if you want to drive lasting, meaningful behavior change, then haptics could be working against those goals.

This is because most safety devices with haptic feedback produce adverse stimuli. It only activates when there's a perceived problem, making haptics a type of punishment. Often incorrectly referred to as negative reinforcement, which would be the removal of adverse stimuli, instead the delivery of adverse stimuli is appropriately referred to as punishment because of the psychological effect it has. The "buzz" is delivered as a negative consequence in an attempt to condition the worker and shape their behavior as they are automatically targeted as the problem needing to be solved. Furthermore, this approach does nothing to reward or praise users for good behavior. Although leaders often like the idea of an easy and automatic way to improve mindfulness, neither of these aspects contribute to long-term change, which is the ultimate goal of improving any safety culture.

HAPTICS – GREAT FOR CONSUMERS, BUT NOT NECESSARILY SAFETY

To understand appropriate applications for haptic technology, it's helpful to recognize the purpose of this technology. The goal of vibrotactile motion is real-time feedback. A device typically delivers gentle force to the wearer through vibrations, jolts, or buzzes. The logic behind this technology is that since humans aren't always in a position to see or hear a cue, a harmless physical force is a suitable way to get their attention.

A popular example is the Apple Watch, which included its Taptic Engine when first released in 2014 ("Apple Unveils Apple Watch"). For the past decade, users have felt a slight buzz whenever they receive a notification. The feature then evolved from alerts and keyboard impressions into gaming experiences and meditation apps ("Apple Watch").

In the health category, wearers of fitness trackers like FitBit, Garmin, and Amazfit experience nudges to get moving. Similar tactile feedback can be found in sit-stand

DOI: 10.1201/9781003432302-15

FIGURE 12.1 Chapter 12 sidebar QR operant-conditioning. SIDEBAR: Video – Check out these fun clips from TV's The Big Bang Theory, where negative reinforcement vs. punishment is debated, and effective behavior shaping through positive reinforcement is explored. https://makusafe.com/blog/operant-conditioning/

desks, ergonomic office chairs, and back braces. Users experience a pulse whenever it's time to change positions or poor posture is detected. Tesla even has a full-body suit that "provides touch and force feedback to train and develop an athlete's muscle memory," including the ability to reduce "the risk of injury during intense training sessions ("New Level of Sports Training").

These consumer examples have one thing in common – the haptic feedback isn't punitive. It delivers neutral communication (you have a message) or a positive celebration (you hit your step count), with no intention to produce guilt or a "gotcha" moment. Even when the buzzing is a reminder, like it's time to drink more water, it's a low-stakes situation. Missing a movement goal for the day won't lead to serious consequences or an immediate health crisis.

However, haptic feedback in industrial environments has an entirely different context. First, a life-or-death situation is a reality on a manufacturing floor or assembly line. There are threats that can have a dire effect on workers: being caught in, caught between, or struck by machinery, and the likes. Ignoring or not understanding a safety cue is a major risk for an injury or a fatality.

Second, haptic feedback as a safety alert is most often a negative message – it's not usually neutral or positive. There's a vibration or buzz only when something is flagged as a risky action, movement, or threat. The haptic feedback's sole function is to deliver a warning.

Why is haptic feedback a problem when a factory is already filled with alarm systems? For the simple reason that those notifications are a form of mass communication. When they deploy, they are directed at everyone – no individual is called out or targeted. Furthermore, not a single version of those alarms is attached to a human body.

DIFFERENCES BETWEEN ALARM SYSTEMS AND HAPTIC FEEDBACK

Alarm Systems	Haptic Feedback
Facility-wide communication	Individual notification
Mounted on a wall, ceiling, or equipment	Worn directly on a worker's body
Uses sound or light	Delivers buzzes or vibrations

THE DOWNSIDE OF HAPTIC FEEDBACK

Another reason why haptic technology may be counterproductive as a safety warning is the human factor. What goes missing from the discussion is what happens to people when they are constantly subjected to negative consequences or punishment. They begin to equate the sensations with uncertainty, a nuisance, an interruption, or a reprimand.

1. ***Poor Context*** – or *Why on Earth Am I Getting Buzzed Again?* Not all haptic communication provides a clear context. Some devices, such as the wearable we developed at MākuSafe, monitor a wide array of potential hazards. But if we delivered haptic feedback, the same or similar sensation might mean numerous things and is open to interpretation. An often used example of a simple signal being interpreted differently might be a flashing yellow light at an intersection. Does that mean stop, slow down, proceed cautiously, or what exactly? Wearers wouldn't have an immediate way to distinguish which reading was triggered. They'll wonder, "Is it because this area is too hot, fumes were detected, I lifted incorrectly, or I nearly tripped?" If they don't know what the issue is, they can't effectively and quickly determine what exactly to do about it. Any hesitation alone could be a safety issue in a situation where time is of the essence. There are some wearable technologies that buzz multiple times depending on what is detected, possibly adding to the confusion.

   A lack of training can also create a knowledge gap. Let's say a company implements a back harness that vibrates when there is risky motion. They even provide an orientation to show how to wear it, emphasizing the tool is for protection. But they don't talk about the big picture of how many workers are suffering from strains or why it's important for everyone to avoid injuries. Nor do they teach employees about how to respond to a haptic warning, like reviewing proper lifting techniques. Employees may not understand why a behavior change is necessary in the first place, much less how to make corrections in the moment or retrain their bodies to move safely. Further, the worker's behavior may not be the issue at all. They likely know this. An adjustment to a process or elevated workstation may eliminate the need for a particular motion altogether.
2. ***Easy to Ignore*** – or *This is Just As Annoying As My Snooze Button.* Humans are incredibly good at tuning out unwanted interference. Think of how easy it is to disregard pings from chat platforms. Notification desensitization, or alert fatigue, is a real concern. In fact, studies have shown that even nurses in intensive care units can become overwhelmed by and desensitized to alerts, ultimately "interfering with caring for patients and causing reduced trust in alarm systems" (Lewandowska et al.). Especially if workers don't see the value or purpose of haptic sensations, they are likely to stop paying attention to them.

   Imagine wearing a gas monitor in a confined space where fumes are known to be present but not usually at a dangerous concentration. While the device is a sensible precaution, what happens if the tactile sensors are set to vibrate regardless of the reading level? The worker is repeatedly buzzed with what amounts to false alarms. At best, that's mildly annoying and possibly distracting. At worst, the employee disregards a valid reading and puts themselves and possibly others in harm's way.

3. ***Distraction*** – *Now My Concentration Is Broken and I Messed Up*. Haptic feedback is intrusive and disruptive by design. Workers have no idea when they are going to receive a pulse. That inherent randomness doesn't make sense in settings where people need their focus to stay safe.

   The risk of a startle reflex, no matter how minor, can have serious ramifications. Like a welder who has to restart an advanced weave bead, setting back a project's timeline. A food packaging worker who accidentally drops foreign material into a batch, escalating into a costly recall and a PR disaster. A forklift driver who is buzzed when pedestrians are nearby slams on their brakes, and the sudden movement causes the load to be dropped. In some environments, or in the case of a load of hazardous material load, would it be better to alert the pedestrian? Not to mention all the situations where someone unexpectedly jerking could lead to a lost finger or worse.

4. ***Punishment*** – or *I'm In Trouble Again*. There's also a reality that haptic feedback will make users feel like they are receiving a "wag of the finger." If workers feel like they are wearing a version of Big Brother, they'll form an antagonistic relationship with these tools. They'll even associate the delivery of negative consequences with the company rather than the technology, only adding to the negative perception.

   Let's return to the haptic back harness. What happens if the device is issued with the assumption that workers are to blame for improper lifting? For starters, management won't pause to review if there are environmental issues like an obstructed shelf or bulky materials that are causing a team to struggle with ergonomics. Second, they won't ask their workforce for suggestions on how to make this activity easier, like vacuum lifters or lift tables. The haptic feedback is only going to breed resentment because it's not addressing the actual safety issue.

Here's the end result with any four of these scenarios – workers will find a way to tune out, avoid wearing, or disable the haptic feedback. They might even learn how to move or work in a way that avoids the buzzer, ultimately putting themselves in a riskier position. None of this is going to result in long-term safety gains.

REINFORCEMENT THEORY AND SAFETY MANAGEMENT

Let's step away from haptic technology for a moment and into the world of change management and behavior theory. Behaviorism examines how people learn through conditioning. It is especially interested in how individuals are affected by positive or negative elements in their surroundings. One of the classic expressions of this concept is Pavlov and his conditioning with salivating dogs.

As this school of thought progressed, B.F. Skinner added his theory – that people are primarily controlled by their environment and thus their behavior can be controlled to produce a specific outcome ("B. F. Skinner"). Whereas Pavlov used unconscious conditioning, associating the stimulus of a metronome with feeding times, Skinner advocated for "the rewarding of a partial behavior or a random act that approaches the desired behavior" ("B.F. Skinner: 1904–1990"). He further believed that "everything we do and are is shaped by our experience of punishment and reward" ("B.F. Skinner: 1904–1990").

What he found, and many others have explored since, is that there is a fundamental difference between positive reinforcement and the delivery of negative consequences, aka punishment. If the consequences to an action are interpreted as a reward rather than punishment, the desired behavior is more likely to begin to take root as a habit. Penalties only have limited success at producing lasting change. Parents often find this with children – a rebuke can stop someone in their tracks and achieve immediate compliance. But even with adults, if you want someone to permanently adopt a different behavior, a series of rewards is more motivating.

This is the nexus where haptic biofeedback has been misapplied. In consumer devices, the gentle pressure is usually neutral or positive and used to reinforce a beneficial habit. But a safety device with haptics does neither – it delivers negative stimuli as a way to tell a worker to stop something.

Furthermore, haptic feedback does nothing to address the root of many safety issues – an environmental hazard. It once again assumes that the worker is the root problem, not something in their surroundings. It also assumes that the worker knows what to do to mitigate the hazard or reduce the risk. Haptics alone can't replace the insights and impacts employed by EHS professionals, ergonomists, and industrial hygienists. While it can call attention to a risk, that's only the first step. It is not a substitute for the benefits of conducting a safety analysis, implementing corrective measures, increasing training, or reengineering work to minimize hazards.

Haptic feedback may seem like an "easy button" for promoting better safety behavior, but at best it's a short-term fix to a more important underlying problem. To sustain an effective safety culture, companies need to use positive reinforcement instead. This can only be supported by a robust ecosystem of incentives, praise, recognition, communication, technology, and reengineered work.

KEY TAKEAWAYS

- Can be distracting, confusing, or even frustrating to workers
- Does not educate, encourage, or empower
- Ultimately not effective at creating permanent, meaningful change

BIBLIOGRAPHY

"Apple Unveils Apple Watch—Apple's Most Personal Device Ever." *Apple*, 9 September 2014, https://www.apple.com/newsroom/2014/09/09Apple-Unveils-Apple-Watch-Apples-Most-Personal-Device-Ever/

"Apple Watch User Guide." *Apple*,2024, https://support.apple.com/en-au/guide/watch/apd371dfe3d7/watchos

"B. F. Skinner." *Harvard University – Department of Psychology*, 2024, https://psychology.fas.harvard.edu/people/b-f-skinner

"B.F. Skinner: 1904–1990." https://www.pbs.org/wgbh/aso/databank/entries/bhskin.html

Lewandowska, Katarzyna, et al. "Impact of Alarm Fatigue on the Work of Nurses in an Intensive Care Environment—A Systematic Review." *International Journal of Environmental Research and Public Health*, vol. 17, No 22, 13 November 2020. *National Library of Medicine*, https://www.ncbi.nlm.nih.gov/pmc/articles/PMC7697990/

"New Level of Sports Training with Haptic Technology." *TESLASUIT*, 9 October 2023, https://teslasuit.io/blog/sports-training-with-haptic-technology/

13 How Wearables Contribute to the Prevention of Serious Injuries and Fatalities

In any industry, the most valuable asset is its people. The well-being of employees is not only a moral imperative but also a critical factor in the sustained success and reputation of an organization. It is wise for organizations and safety professionals to focus on preventing life-altering serious injuries and fatalities (SIFs) by concentrating on high-hazard types of work. These companies may consider moving to ASTM E2920 as a standard for recording injuries based on severity of incidents and meaningful connection to work. Often we hear that because they and their organizations are committed to an established strategy for focusing on SIFs and potential serious injuries and fatalities (P-SIFs), they don't see how wearable sensor tech will help them with their priorities, major incidents, or saving lives.

Yet research by the Campbell Institute of the National Safety Council pointed to the need for gathering and analyzing data as crucial. They recommended observations, walkarounds, and conversations as important to revealing precursors to SIF ("Designing Strategy for Serious Injury and Fatality Prevention"). We believe using wearable sensor technology to understand the context of work can have a profound impact on improving safety outcomes. Especially when paired with advanced push-to-talk communication capabilities, companies have an opportunity to understand precursors and gain the perspective of the frontline workers, which will contribute to SIF prevention.

THE VALUE OF WORKER INPUT FOR SIFs AND PSIFs

Certain types of work are inherently more hazardous and thus are primary focus areas for SIF prevention efforts. Along with vehicles and roadway work, these types of work are responsible for the most severe incidents and loss of life across industries.

Hazard	Description
Confined Space Entry	Working in spaces with limited entry and exit, such as tanks, silos, and storage bins, especially where hazardous atmospheres or other dangers may be present.
Vehicle Safety	Operation and maintenance of vehicles, including company cars, trucks, and heavy machinery, emphasizing safe driving practices and vehicle condition.

DOI: 10.1201/9781003432302-16

Hazard	Description
Electrical Work	Tasks involving electrical systems and equipment, where the risk of electric shock, arc flash, and fire are significant concerns.
Forklifts and Mobile Powered Equipment	Safe operation of forklifts, cranes, and other mobile machinery to mitigate collisions, tip-overs, and other accidents.
Lockout/Tagout (LOTO) Energy Isolation	Procedures ensuring machinery is properly shut off, de-energized, and unable to start up again prior to the completion of maintenance or repair work.
Hazardous Materials	Handling, storing, and disposing of chemicals and other hazardous substances, where exposure can lead to serious health risks.
Working at Height	Tasks performed at elevated levels, such as on ladders, scaffolds, or rooftops, where falls are a high risk.
Welding and Hot Work	Activities that produce heat, sparks, or flame, such as welding, cutting, and brazing, posing risks of fire and burns.
Excavation and Trench Work	Digging and working in trenches or other excavated areas, where cave-ins and other soil-related hazards are a concern.
Bulk Material Handling and Storage	Movement and storage of large quantities of materials, where the risks include spills, equipment failures, and manual handling injuries.

These workers are the first to encounter potential hazards in their daily tasks. Their direct interaction with machinery, materials, and processes means they are acutely aware of the practical challenges and dangers inherent in their work. This firsthand experience enables them to identify risks that may not be evident to managers or safety officers who are not present in the field.

For instance, a worker operating heavy machinery might notice that a particular safety guard is prone to malfunction under certain conditions. Reporting this observation can prompt a timely intervention that prevents a potential SIF. Without the worker's input, such an issue might only be discovered after an incident occurs, potentially resulting in injury or fatality.

One of the most memorable examples we've heard about was a worker charged with identifying hazards on a ship. He happened to notice boot prints on top of a pipe right next to the railing. It was uncovered that other team members were routinely climbing on top of this elevated pipe so they could reach equipment up high, forcing them to hang over the rail while under way at sea. The potential for someone to go overboard and be lost was extraordinary, yet this risk was only identified because of a boot print.

This perfectly illustrates why preventing SIFs involves not only reacting to incidents but also learning from near-misses and seeking out where serious injuries or fatalities could occur. Workers' insights are crucial in these scenarios. By listening, learning, empowering, and analyzing P-SIF events, organizations can identify underlying hazards and systemic issues that need to be addressed.

For example, a near-miss involving a falling object might reveal weaknesses in a company's material handling procedures. Workers who witness or are involved in such incidents can provide detailed accounts of what occurred, why it happened, and how similar events can be avoided. This feedback loop is essential for continuous improvement in safety practices.

THE ADVANTAGES AND IMPACT OF PUSH-TO-TALK VOICE REPORTING

In recent years, the advent of wearable technology with push-to-talk (PTT) voice communication capabilities has revolutionized the way workers can contribute to safety efforts. This capability, which we chose to include in our MākuSafe device, offers several advantages that enhance the quantity and quality of worker input, leading to better prevention of SIFs and a deeper understanding of P-SIFs.

PTT functionality allows workers to report hazards and incidents in real time, especially in dynamic work environments where conditions can change rapidly. This immediacy ensures that safety issues are documented and addressed promptly, reducing the lag time between hazard identification and mitigation. This elevates the voice of the frontline worker, breaks down barriers, and aids inclusion of needed perspective.

For example, a construction worker on a large multi-contractor job site finds an alarming structural issue. Using the PTT voice communication, they record the following message: "You better come down and see this I-beam – you are not going to like what I found!" The safety manager receives a real-time notification about the report, takes 30 seconds to listen, and immediately arrives to investigate. After confirming the worker's observation, they issue a stop work order until the problem is remedied.

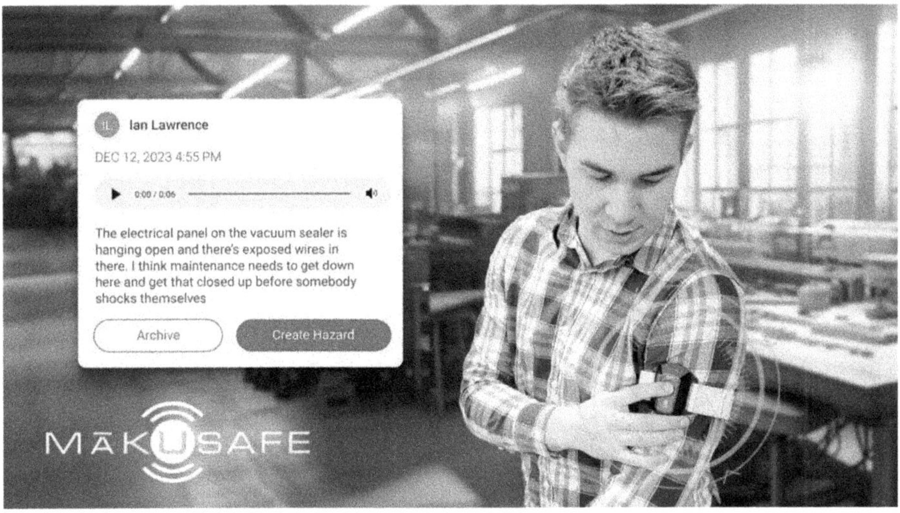

FIGURE 13.1 A worker reporting an electrical hazard using push-to-talk voice communication. Only a few seconds of their time is needed, but their observations can make a critical difference.

The simplicity of PTT devices also encourages more frequent and spontaneous reporting. Workers can quickly relay information without needing to stop their work to fill out forms or access a computer. This ease of use lowers the barrier to reporting, leading to higher engagement and more comprehensive safety data collection.

CONTEXTUAL DATA ON WORKING EXPERIENCES AND EXPOSURES

Wearable sensor devices can provide data about the context in which work is being performed. This data includes variables such as heat index, sound exposure, air quality, physical exertion levels, repetitive movement, location, time on shift, task being performed, and equipment being used. Coupled with input and observations from PTT communication, leaders can begin identifying patterns and developing targeted interventions. Data can be integrated with safety management systems to automatically log reports and track trends over time. This integration provides a rich dataset for analyzing safety issues. This centralized intelligence can enhance predictive analytics, helping organizations to anticipate SIFs more effectively.

Incorporating this contextual data offers numerous benefits:

Improved Learning from Near-Misses

Contextual data provides valuable insights analyzing and learning from near-misses. Understanding the exact conditions under which an incident occurred helps identify

FIGURE 13.2 Wearable sensor data can provide insights into working conditions and how they might impact safety, such as heat, noise, air quality, and low light.

root causes and contributing factors, leading to more effective corrective actions. For instance, knowing that a worker was performing a high-risk task with inadequate equipment and was affected by environmental conditions at the time of a near-miss can prompt a review and improvement of task-specific protocols and equipment standards.

Comprehensive Risk Assessment

Contextual data enables a more comprehensive risk assessment by providing a detailed understanding of the environmental and operational factors contributing to safety hazards. For example, high heat index readings combined with prolonged physical exertion can indicate an increased risk of heat-related illnesses. Addressing these conditions proactively can prevent incidents.

Personalized Safety Measures

Data on individual worker exposures and conditions can help tailor safety measures to specific needs. If certain workers are consistently exposed to high noise levels, targeted interventions such as providing enhanced hearing protection or rotating tasks can be implemented to reduce their risk of hearing loss.

Enhanced Predictive Analytics

Collecting and analyzing contextual data enhances predictive analytics capabilities. By identifying patterns and correlations between various factors and safety incidents, organizations can predict and mitigate risks more effectively. For example, tracking the correlation between air quality and respiratory issues can lead to improved ventilation systems in specific areas.

Better Workforce Management

Data on factors such as time on shift and physical exertion levels can inform workforce management decisions. Monitoring these parameters helps prevent worker fatigue and overexertion, reducing the likelihood of errors and accidents. Ensuring adequate rest periods and task rotation can enhance overall safety and productivity.

THE IMPACT ON SAFETY OUTCOMES

When worker input is systematically incorporated into safety management systems through advanced technologies like wearable PTT devices and contextual data collection, the impact on safety outcomes can be profound. Organizations that prioritize worker feedback tend to have lower rates of SIFs and a stronger safety culture. Workers feel empowered to take ownership of their safety and that of their colleagues, leading to more vigilant and conscientious behavior on the job.

Moreover, the data collected from worker reports and contextual monitoring can be used to track safety trends, identify recurring issues, and develop targeted interventions. This data-driven approach ensures that safety initiatives are based on real-world conditions and experiences, making them more effective and relevant.

KEY TAKEAWAYS

- Workers provide critical insights that are essential for identifying precursors and contributing factors to SIF events.
- Push-to-talk voice reporting via wearable devices, along with the collection of contextual data about working experiences and exposures via sensors, aids learning and analysis to uncover where potential for SIFs is present.
- By establishing robust feedback mechanisms and valuing the contributions of workers, organizations can significantly enhance their safety outcomes.

BIBLIOGRAPHY

"Designing Strategy for Serious Injury and Fatality Prevention." *Campbell Institute*, 2020, https://www.thecampbellinstitute.org/wp-content/uploads/2020/02/Campbell-Institute_Designing-Strategy-for-Serious-Injury-and-Fatality-Prevention.pdf

14 The New View of Safety

The roots of new approaches to safety can be traced back to Paul O'Neill's tenure at Alcoa in the late 1980s and early 1990s. O'Neill's revolutionary focus on worker safety not only transformed Alcoa but also set a new standard for safety management in the corporate world. Many of the professionals who were part of this transformation at Alcoa went on to influence other organizations, contributing to the evolution of safety management that brings us to where we are today.

Several concepts have emerged in recent years that embody this new approach to safety: Human and Organizational Performance (HOP), Safety 2.0 by Erik Hollnagel, Safety Differently from Sydney Dekker, and the New View of Safety. While these concepts have some differences, they share a common foundation: the application of modern management principles to safety practices.

HUMAN AND ORGANIZATIONAL PERFORMANCE (HOP)

HOP is attributed to thought leaders such as Todd Conklin, Bob Edwards, Rob Fisher, and others. It represents a significant shift from traditional safety management approaches. The willingness of these thought leaders to share these concepts and their work openly in the hope of widespread adoption and improved results is appreciated and noble.

The National Safety Council appropriately points out that HOP is science based and risk focused. They define HOP not as a safety and health discipline but as "an operating philosophy combining human factors engineering with organizational psychology and leads to better, more efficient, more resilient and safer workplaces where employees and leaders are both engaged and valued by each other and the organization" (Human & Organizational Performance). Its core principles include the following.

Error Is Normal

People make mistakes, and even the best people do. Error isn't a choice. Ever have your keys in one hand and trash in the other but have to remind yourself which one gets tossed in the bin in front of you? The vast majority of the time, work is successful and mistakes don't have negative outcomes.

Workers Are Adaptive Problem Solvers

They routinely find ways to solve problems and adapt to get work done successfully. This is because job instructions and SOPs never completely address every possible scenario and nuance.

DOI: 10.1201/9781003432302-17

Context Influences Behavior while Systems Drive Outcomes

Organizations put people in the position of having to make adjustments and decisions about the right thing or an acceptable way, every day. You should think of your frontline workers as real-time risk managers.

Our Response to Failure Matters

Blame and punishment fix nothing. The traditional response of being punitive hampers any learning and further destroys any chance for open and transparent communication. Blame is the enemy of improvement.

Learning and Improvement Are Vital

Learning is deliberate and doesn't happen by accident. You can punish or choose to learn, but it's not possible to do both effectively. Consider the difference between an accident investigation and an event analysis. Organizations that embrace learning benefit greatly. In the context of HOP and safety management, many will implement well-designed organizational learning teams, events, and methodologies.

Imagine for a moment that a company fired an employee if they made a mistake or strayed from a SOP. That company would learn nothing about why their employee was compelled to do the work differently than designed. It's also likely the next person they put in that role will find themselves unemployed when they encounter the same conditions. The cost of recruiting and training for that role would be extremely high.

PERSPECTIVE FROM THE INSURANCE INDUSTRY

The evolution of safety management is not occurring in isolation. The insurance industry, particularly in workers' compensation, is undergoing a transformational shift. Forward-thinking leaders in insurance are no longer content to just issue payments to repair or replace losses. Instead, they are seeking to engage with policyholders proactively to help predict and prevent losses before claims occur.

This shift in the insurance industry aligns perfectly with the new approaches to safety management. Companies like MākuSafe, which is based in one of the top insurance hubs in the United States, are well positioned to bridge the gap between these new safety philosophies and the evolving needs of the insurance industry.

LEARNING FROM NORMAL WORK

One common question from organizations exploring wearable safety technology is: "We don't have many incidents or we've already reduced our incident rate significantly – why would we use your wearable solution? What else is there for us to learn?"

This question highlights a crucial aspect of the new view of safety: the importance of learning from normal work. Erik Hollnagel suggests that instead of aiming for "zero incidents" as a goal, we should establish "work to 100%" as an objective. This approach allows organizations to relentlessly develop resilience potential in all aspects of work, even when incident rates are low. Recalling the Blue Line Model,

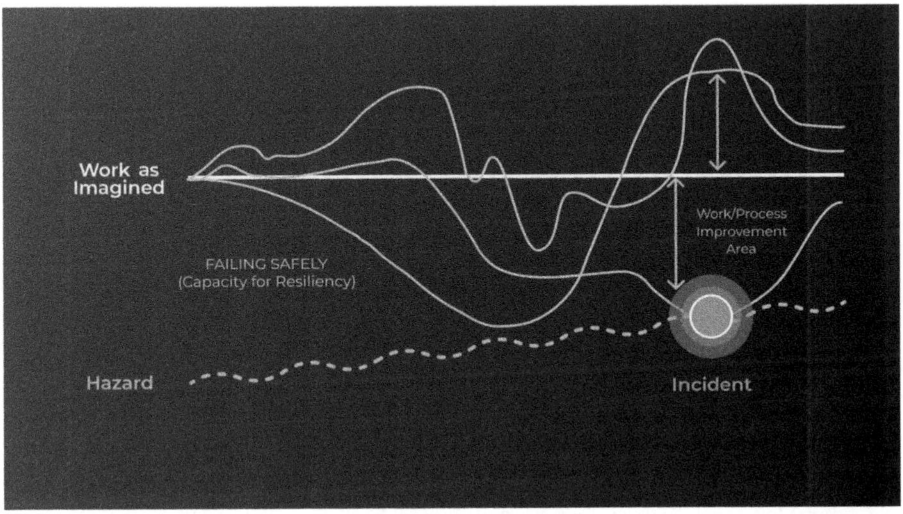

FIGURE 14.1 Work as performed is represented here by the blue lines in the model. On good and bad days alike, the realities of how work gets performed do not follow a straight path as it was imagined or designed. Workers are forced to adapt to circumstances as they arise. By truly understanding the realities of how work is performed, we can improve work processes and create capacity to work safely, even when systems fail or people inevitably make errors.

consider the value of implementing practices like Blue Line Reviews or Learning Team Meetings.

Learning from normal work is essential because:

1. It helps understand how work is actually done, not just how it's imagined or prescribed.
2. It reveals successful adaptations and best practices that can be shared across the organization.
3. It identifies potential risks before they lead to incidents.
4. It builds a culture of continuous improvement and learning.

This is vital because today's work environments are complex systems with numerous interacting components. As Todd Conklin aptly put it, the goal should be to "Fix the work, not the people." This shift in focus from individual behavior to systemic factors is a key principle of the new safety paradigm.

Implementing these new approaches to safety requires a high degree of emotional intelligence, as described by Daniel Goleman. Leaders must be able to understand and manage their own emotions and those of others to create a culture of openness and learning.

Moreover, the concept of humble inquiry, introduced by Edgar Schein, is crucial for organizational learning. The ability to ask good questions of those who do the work every day and gain meaningful insights is essential for leaders. This approach leads to a culture of trust and continuous improvement.

THE ROLE OF TECHNOLOGY IN THE NEW SAFETY PARADIGM

Wearable sensor technology, like that developed by MākuSafe, is particularly well suited to support and accelerate these new safety initiatives. By providing real-time data on environmental conditions, worker movements, and potential hazards, wearable technology enables organizations to:

1. Normalize and anticipate mistakes and provide opportunities for learning without blame.
2. Shift focus from blaming individuals to understanding systemic factors.
3. Enable continuous learning and improvement to build resilience and the capacity to fail safely.
4. Provide context about the work environment, helping understand how systems drive behaviors and outcomes.
5. Allow for immediate and informed responses to potential safety issues.

For example, a production line worker learns her coworker is regularly getting a bruise on her arm from a certain part of a machine. She can use the voice communication feature on her wearable to recommend adding padding to the machine. The safety manager can then investigate the situation further to learn how widespread this issue is among workers. Listening to this team also uncovers a process improvement suggestion, which supports quality standards and eliminates an awkward lifting motion.

As technology facilitates communication and breaks down barriers, it creates opportunities for positive conversations focused on risk in the work. This directly leads to a more positive culture by creating engagement and participation.

The new view of safety, embodied in approaches like HOP and supported by advanced technologies, represents a significant shift in how organizations manage safety. By focusing on understanding human factors, system influences, and the importance of learning from normal work, these approaches promise to create safer workplaces and more resilient organizations. Technology with the right approach can be very human centric and is in alignment with contemporary safety management philosophy.

KEY TAKEAWAYS

- The integration of new safety philosophies with emerging technologies and evolving insurance practices offers exciting possibilities for the future of workplace safety.
- Organizations that embrace these approaches and leverage advanced technologies will be well positioned not only to manage safety but also to manage safely in an increasingly complex world.

BIBLIOGRAPHY

"Human & Organizational Performance: A path to improvement for all organizations." *National Safety Council*, 2021, https://www.nsc.org/getmedia/6dc7b76d-09a2-4ad5-9be2-0b0365c34b46/nsm-wk4-mi

15 How Organizational Culture Shapes Innovation Adoption

In the ever-evolving landscape of workplace safety, the introduction of innovative technologies holds the promise of safeguarding workers, reducing incidents, and potentially saving lives. However, the successful deployment of such technologies, particularly wearable safety devices, is not solely dependent on their potential benefits. Instead, an organization's existing culture and leadership play a pivotal role in either enabling or limiting the adoption of these innovations, particularly when it involves asking employees to don a wearable device every day. The thoughtful implementation of wearable technology can also contribute to building stronger safety cultures.

THE CULTURE ICEBERG: VISIBLE AND HIDDEN INFLUENCES

To understand the impact of organizational culture on innovation adoption, we must first explore the concept of culture itself. The iceberg model provides an apt analogy for organizational culture, with observable elements above the surface and hidden yet influential factors beneath.

Above the waterline are the tangible aspects and behaviors of an organization:

- How people treat each other
- Physical workspace layout, neatness, and organization
- Dress codes
- Company logos and slogans
- Visible safety measures and signage
- People speaking up and actively contributing
- The willingness of workers to lead or follow

However, it's what lies beneath the surface that truly shapes what you see when you walk around:

- Deeply held attitudes, beliefs, and perceptions
- Shared values and assumptions
- Unwritten rules and norms
- Historical experiences and collective memory
- Power dynamics and relationships

DOI: 10.1201/9781003432302-18

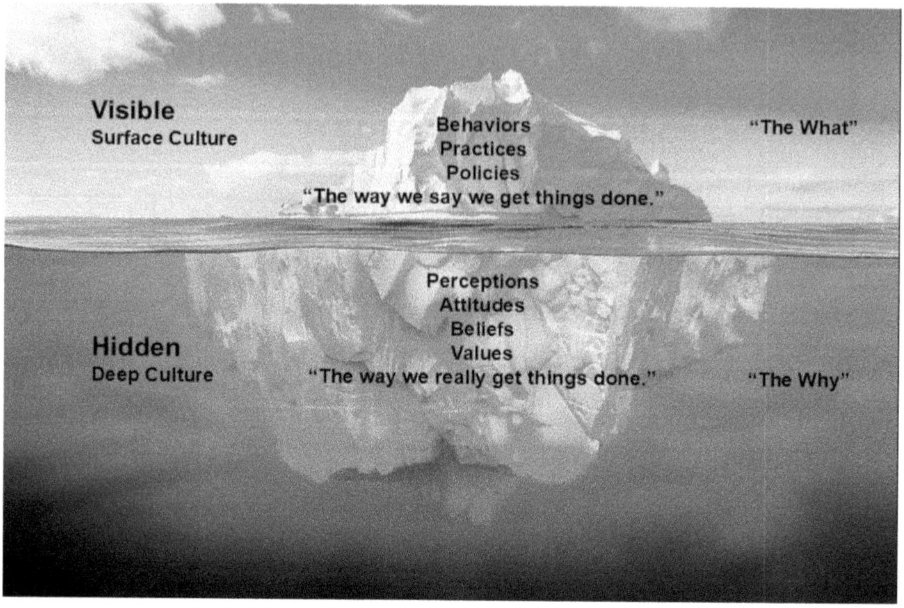

FIGURE 15.1 Workplace culture is like an iceberg – only the most obvious elements are visible. The majority, however, are more subtle and often unwritten. These "silent" cultural marks are what predominantly drives behavior.

These hidden elements form the bedrock of an organization's culture, influencing how employees interpret and respond to new initiatives, including the introduction of wearable safety technology.

LEADERSHIP IS THE CATALYST FOR CULTURAL CHANGE

When discussing organizational culture in the context of safety, the term "safety culture" often emerges. However, this concept has been a subject of debate among scholars and practitioners. The reality is that safety culture is not a separate entity – it is an integral part of the overall organizational culture. It reflects how safety is prioritized, communicated, and integrated into daily operations and decision-making processes.

The debate surrounding safety culture stems from the difficulty in isolating it as a distinct element. Instead, it's more accurate to view safety as a core value that permeates all aspects of organizational culture, shaping behaviors and decisions at every level.

In navigating the cultural landscape, leadership plays a crucial role. Effective leadership is critical in inspiring and shaping how things are done, setting the tone for how employees generally think, and creating alignment between personal views and the organization's values. One compelling perspective on leadership asserts that it involves "inspiring people to see a vision of what can be in such a powerful way that they choose to follow you there." This approach is particularly relevant when introducing innovative safety technologies like wearable devices.

However, leaders face significant challenges in this endeavor. More often than not, they have to start by overcoming the "Big Brother" myth. Employees often have immediate concerns about wearable technology, including:

- Will my employer now know my medical and personal information?
- Is this going to keep buzzing me and interrupting my work?
- Will the company know where I'm at all the time?
- Is the company just trying to find performance reasons to get me in trouble or get rid of me?

To address these concerns, leaders must clearly communicate the limitations and actual capabilities of the technology. They need to be willing and able to persuade workers to consider an alternate viewpoint. For instance, we at MākuSafe have gone to great lengths to develop a wearable safety technology so that we would not:

- Monitor biometrics, health data, or PII
- Constantly track employee location
- Know when someone is on break or how many times they visit the restroom
- Listen to worker conversations

We often point out that the data we gather about heat levels, sound octaves, or slip/trip motion should be viewed as a way to hold leaders accountable for fixing hazards – not anything that could be used against a worker. In organizations with adversarial relationships between management and labor, inspiring a shared vision becomes

FIGURE 15.2 Without a strong culture of trust and respect or leadership commitment and influence, the reaction of workers to wearable tech may be met with distrust and cynicism.

exponentially more difficult. Leaders must work to build trust and open communication channels before introducing potentially controversial technologies. Leaders will need to articulate how the technology aligns with the organization's values and how it benefits both the company and individual employees.

While organizational culture influences technology adoption, the relationship is reciprocal. Technologies like MākuSafe can, in turn, shape and enhance organizational culture when implemented thoughtfully. MākuSafe's MyVoice communication feature can elevate frontline worker perspectives if leaders genuinely value their input. This creates opportunities for meaningful conversations, allowing leaders to seek in-depth employee comments and suggestions, explore needs and concerns, and express gratitude for insights shared. These interactions can foster new behavioral habits, resulting in higher engagement and participation in safety programs. Over time, this can lead to a more robust safety culture and a general improvement in organizational communication and trust.

For example, a safety leader notices readings for high TVOCs and CO_2 detected. They initiate a discussion between the shift supervisor and workers, which reveals batches of raw materials from a supplier that smell different. Together, they created a protocol to leave a voice report whenever materials arrive that smell unusual. This back-and-forth communication allows the company to better track quality concerns through production as well as address any supplier issues.

THE DECISION-MAKING NEXUS

Ultimately, an organization's ability to keep people free from harm, hazards, and risks stems from decision-making throughout the entire organization and all of its interrelated parts. These decisions are fundamentally based on the system of shared values within the organization. When safety is deeply ingrained in the organizational culture, it influences decisions at all levels, from strategic planning to daily operations. This holistic approach ensures that safety considerations are not an afterthought but an integral part of every decision, reducing the likelihood of unintended risks or consequences.

Imagine a worker who can use voice reporting because they note the new flooring throughout their entire department is slick. Triggered by this timely observation, the safety manager can evaluate the situation. They might find a hidden issue, like the wrong paint was used. The maintenance team mistakenly used a gloss paint with no added grit, which now poses a slip or fall risk. Safety and maintenance can then work together to remediate this issue – all because technology and a strong safety culture empowered the worker to record an invaluable 30-second message.

As we look ahead, it's clear that the use of wearables and other innovative safety technologies will become increasingly ubiquitous, expected, and integrated into how work is performed. However, because we are still in the early stages of adoption, there are challenges to successful deployment that companies will need to consider and address. Organizations with positive cultures and engaged workers are better positioned to overcome signs of resistance or hesitation. They have the foundation necessary to address concerns, build trust, and effectively implement new technologies.

KEY TAKEAWAYS

- Only a portion of workplace culture is easy to observe. When strengthening safety attitudes, organizations will need to leverage those that are hidden undercurrents.
- While innovative safety technologies like MākuSafe offer tremendous potential to improve workplace safety, their successful adoption is inextricably linked to an organization's existing culture and leadership.
- By recognizing this relationship and actively working to create a culture that values safety, innovation, and employee input, organizations can pave the way for successful technology adoption.

16 Management Effectiveness, Systems Orientation, and Certifications

In the realm of organizational management, a systemic approach is crucial for achieving both effectiveness (goal attainment) and efficiency (wise resource use). This systems orientation views management as a continuous process, ensuring that organizations can adapt, improve, and thrive in dynamic environments. We believe strongly that wearable tech can strengthen the safety management process, improve outcomes, and provide focus for leaders.

THE MANAGEMENT PROCESS

While various models may differ slightly in their steps, the four core functions of the management process consistently include:

Planning – Where Do We Want to Go?

Planning is the foundational step in any management process. It involves setting objectives and determining the best course of action to achieve them. Despite its critical role, planning is often neglected, resulting in unclear goals and directions. Without proper planning, it is very difficult to understand if the plan's objectives have been met. It can also be challenging to assess where things went wrong. If there's no clear plan in place, then how will you know where and why an action has deviated from an objective?

Organizing – How Will We Align Our Resources to Get There?

Organizing involves arranging resources – human, financial, and material – in a way that effectively supports the achievement of the goals set during the planning stage. This includes creating structures, roles, and responsibilities.

Leading – Who Will Carry Out the Work?

Leading is about inspiring, motivating, and guiding people to work toward the organization's goals. Effective leadership is built on communication, fosters a positive work environment, and encourages collaboration and commitment.

DOI: 10.1201/9781003432302-19

Controlling – How Will We Monitor and Measure to Ensure Desired Outcomes?

Controlling involves monitoring progress and making adjustments as necessary. It ensures that the organization remains on track to achieve its goals by evaluating performance and implementing corrective actions when needed. Much like planning, evaluating is often overlooked, leading to missed opportunities for improvement. Even when things have seemingly gone according to plan, evaluation helps uncover where members of the team may have taken alternative actions that still resulted in the desired outcome. Knowing that can help with planning for future projects or uncovering spots where you may have just gotten lucky with a decision that wasn't according to plan.

The military uses a process called after-action review (AAR). This process can be easily adapted as a project evaluation in the corporate setting. You can conduct your own internal AAR by asking some of the following questions:

1. What was supposed to happen?
2. What actually happened?
3. What went well? Why?
4. What went poorly? Why?
5. What were the consequences?
6. What are our key learnings?
7. What can we apply to future projects?

It's essential to conceptualize this process as a continuous loop, where feedback fuels ongoing improvement.

SAFETY MANAGEMENT: A PARALLEL PROCESS

The principles of effective management apply equally to safety management. A continuous improvement approach in safety ensures both effectiveness in reducing incidents and efficiency in resource allocation. This systemic view of safety and health management systems (SHMS) has been widely advocated by major associations and thought leadership outlets. The American Society of Safety Professionals (ASSP), the National Safety Council (NSC), and the Occupational Safety and Health Administration (OSHA) all promote the implementation of SHMS. OSHA even provides comprehensive guidelines on safety management systems ("Recommended Practices").

Regardless of the source, the safety management process typically involves common elements including:

1. **Hazard identification:** Identifying potential hazards in the workplace is the first step in preventing accidents and injuries.
2. **Risk analysis:** Assessing the identified hazards to understand the level of risk they pose and prioritizing them accordingly.

3. **Implementing a hierarchy of controls:** Applying controls based on the hierarchy of controls, such as elimination, substitution, engineering controls, administrative controls, and personal protective equipment (PPE).
4. **Evaluating results:** Continuously monitoring and evaluating the effectiveness of the implemented controls to ensure they are working as intended.

Crucially, this process cannot be effective without active participation from frontline workers, a pervasive safety mindset, and strong leadership. These elements are essential for identifying hazards, implementing controls, and continuously improving safety practices.

How this looks in reality might be a shift supervisor who receives a notification of a trip motion detected by his MākuSafe wearable device. Looping in the safety director, they discuss the occurrence with the worker and learn he had been tripping on the small ergonomic mat that's new in his work area. They determine larger mats will pose less of a tripping risk, and a replacement order is made.

SAFETY CERTIFICATIONS: PURSUIT OF EXCELLENCE

Many organizations seek safety certifications as a means of validating their commitment to safety and demonstrating the effectiveness of their safety management systems. This often yields quantifiable positive results. A prime example is Cintas, which became the first company to achieve safety certification at over 100 sites and is now well beyond that (Shaw).

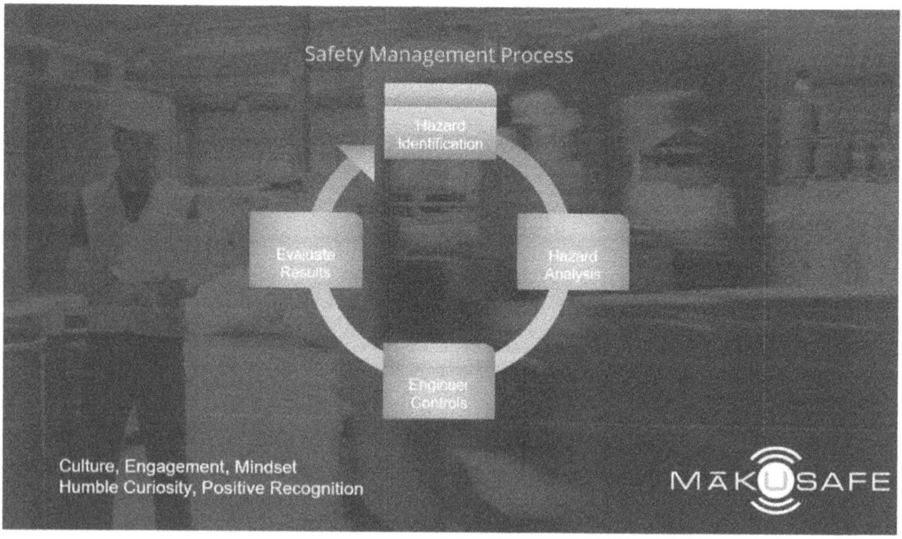

FIGURE 16.1 Improving safety is a continuous loop that cycles between identifying and minimizing risk. Any traditional tendency to view safety management as performing only the first three job functions in a linear fashion and waiting to see what happens is outdated and should be forgotten.

There are various well-recognized alternatives for safety certification, including ANSI Z10, ISO 45001, and VPP Star. While each has its specific requirements, they share common elements that align with general management process steps and basic safety management principles. These include:

1. **Management Leadership:** Commitment from top management to drive the safety program.
2. **Worker Participation:** Involving workers in the safety process to leverage their insights and encourage a culture of safety.
3. **Hazard Identification and Assessment:** Regularly identifying and assessing workplace hazards.
4. **Hazard Prevention and Control:** Implementing measures to prevent and control hazards.
5. **Education and Training:** Providing continuous safety education and training for employees.
6. **Program Evaluation and Improvement:** Regularly evaluating the safety program and making necessary improvements.
7. **Communication and Coordination:** Ensuring effective communication and coordination among all stakeholders, including host employers, contractors, and staffing agencies.

THE ROLE OF TECHNOLOGY IN SAFETY MANAGEMENT

Emerging technologies, such as wearable safety devices, are revolutionizing the field of safety management. Platforms like MākuSafe, when implemented with the right approach, can significantly enhance the effectiveness and efficiency of safety management processes. These technologies can aid in achieving and maintaining certifications by providing:

Automated Hazard Identification: Wearable devices can detect potential hazards in real time, providing immediate alerts to workers and supervisors.

Data-Based Decision-Making: Collecting evidence on workplace conditions helps in making informed decisions to improve safety.

Documentation: Time and date-stamped readings provide a clear record of workplace conditions and incidents. It includes tracking corrective actions and control measures as well.

Human-Centric Reporting: Voice reporting features enable workers to easily report hazards, encouraging greater participation and engagement.

Measurement and Monitoring: Continuous data collection allows for effective monitoring of corrective actions and controls.

Program Improvements: Data insights help in identifying areas for improvement, leading to a more effective safety program.

For example, a worker is responsible for running an industrial power washer and wears ear plugs as hearing protection. However, based on decibel recordings by MākuSafe, it's found that they are routinely experiencing more than the allowable sound dosage within the first couple of hours of his shift. Safety leadership is able to reevaluate hearing protection for this role to ensure protection is commensurate with the exposure level.

By leveraging such technologies, organizations can create a more responsive, proactive, and participative approach to safety management. This not only aids in meeting certification requirements but also supports a culture of continuous improvement in safety practices.

KEY TAKEAWAYS

- Viewing management – and by extension, safety management – as a systemic, continuous process is key to organizational effectiveness and efficiency.
- By embracing this approach, pursuing relevant certifications, and leveraging innovative technologies, organizations can create safer, more productive environments while meeting their strategic objectives.

BIBLIOGRAPHY

"Recommended Practices for Safety and Health Programs." *OSHA*, https://www.osha.gov/safety-management

Shaw, Seyfarth. "Cintas Becomes First Employer to Reach 100 Certified OSHA VPP-Star Worksites." *Seyfarth*, 29 March 2019, https://www.environmentalsafetyupdate.com/2019/03/cintas-becomes-first-employer-to-reach-100-certified-osha-vpp-star-worksites/#:~:text=and%20cost%20savings.%E2%80%9D-,Cintas%20Corporation%20has%20recently%20become%20the%20first%20company%20to%20reach,VPP%20location%20in%20early%202019

17 The Evolution of Management Thinking

Over the past century, our understanding of effective leadership and management has evolved significantly. Early theories on best practices for managing and leading people have been refined and replaced by newer models, each contributing valuable insights into human nature, motivation, and organizational dynamics. Despite this evolution, many of the fundamental challenges in leadership remain. Organizational change efforts often fail, and a significant portion of the workforce remains disengaged, as evidenced by surveys like Gallup's State of the Workforce ("Employee Engagement"). We can all benefit from having a more solid understanding of how to build strong cultures. After all, even world-class athletes practice the fundamentals and have a coach.

This chapter explores the evolution of management thinking, summarizing key concepts and their contributions to today's best practices in leadership. It also highlights contemporary ideas essential for achieving great results with and through people. To successfully introduce new tools and technologies, share visions and values, and expect lasting change, leaders must live and breathe time-tested basics.

"There is no better teacher than history in determining the future. There are answers worth billions of dollars in a $30 history book" – Charlie Munger

"I don't think much of a man who is not wiser today than he was yesterday" – Abraham Lincoln

CLASSICAL APPROACHES

Classical thinking on management still provides valuable lessons. Organizations today cannot operate with the simple carrot-and-stick mentality as they did 100 years ago. Modern organizations must demonstrate concern for people, needs, and satisfaction. However, classical thinkers were some of the originators of those ideas, providing a strong foundation that has evolved over time. Understanding is helpful to today's leaders.

Frederick Winslow Taylor: The Father of Scientific Management

Frederick Winslow Taylor, often called the father of scientific management, was an engineer who introduced a systematic approach to improving efficiency and productivity. In 1909, Taylor published *The Principles of Scientific Management.* He proposed that by optimizing and simplifying jobs, productivity would increase. His key elements included:

- **Harmony and Cooperation:** Emphasized the importance of collaboration between workers and management.

DOI: 10.1201/9781003432302-20

- **Training and Development:** Advocated for the training and development of workers to enhance efficiency.
- **Standardized Work Methods:** Developed scientific methods for performing tasks to ensure consistency and efficiency.

Taylor's focus on systematic training and development laid the groundwork for modern human resource practices and continuous improvement methodologies. However, too much reliance on a scientific approach to productivity without concern for people is outdated and will not work today.

Mary Parker Follett: Transformational Leadership

Mary Parker Follett, rightly recognized as the mother of modern management, was way ahead of her time. Instead of emphasizing a scientific industrial approach, she advocated for what she saw as the far more important human element – regarding people as the most valuable commodity present within any business. In her writings as early as 1927, she introduced the concept of transformational leadership. Her contributions include:

- **Interrelationship of Leadership and Followership:** Emphasized the dynamic relationship between leaders and followers and the power of collective goals.
- **Human Element:** Advocated for the importance of human relations in management, considering people as the most valuable asset in any organization.
- **Positive Human Relations:** Highlighted the need to promote positive human relations to achieve effective management.

Follett's emphasis on the human element and collective goals resonates with contemporary leadership approaches that value employee engagement and empowerment. While this was a shift from what Talyor, Max Weber, and others taught, Follett, Lillian Galbraith, and their contemporaries laid the groundwork for the movement that followed.

Elton Mayo: The Hawthorne Effect

Elton Mayo's experiments at the Hawthorne Works, a Western Electric plant in Cicero, Illinois in the 1920s and 1930s, led to an understanding of what we now refer to as the Hawthorne Effect. Mayo was attempting to discover correlations between lighting levels and productivity, but instead what he found was that in spite of poor environmental conditions, productivity was directly related to how people were treated. Mayo's work underscored the importance of considering human factors in management, paving the way for the human relations movement and modern organizational behavior studies.

The Human Relations Movement and Behaviorism

The era and thinkers that followed paved the way for modern organizational behavior studies and human resource management techniques, emphasizing employee engagement, team building, leadership development, and organizational culture as critical factors for success.

Douglas McGregor: Theory X and Theory Y

Douglas McGregor introduced Theory X and Theory Y in the 1950s and 1960s, exploring assumptions about human motivation and management styles. These beliefs or conclusions not only manifest themselves in a leader's behavior, but have an impact on followers. McGregor depicted a contrast between two opposing ends of a spectrum:

- **Theory X Manager:** Assumes people are inherently lazy and require strict supervision, relies on rules and formal authority, has an "us vs. them" attitude, and has a singular focus on production or outcomes. May achieve short-term compliance out of fear, but likely creates unhappy workers in the long run.
- **Theory Y Manager:** Assumes people are self-motivated and seek responsibility, utilizes encouragement, offers rewards, and strives for satisfaction. Achieves far greater and lasting results with and through people.

McGregor's theories highlighted the impact of management assumptions and leadership style on employee behavior.

Blake and Mouton's Leadership Grid

Blake and Mouton's Leadership Grid evolved from McGregor's work, emphasizing a balanced approach to leadership. Rather than X or Y styles as an either-or choice, the Leadership Grid combined these two dimensions and recognized that each has needed positive elements. Organizations should strive for a high degree of the best that each provides, and neither can or should be sacrificed:

- **Concern for Production:** Focused on achieving high levels of productivity, results orientated, goal attainment, through structure.
- **Concern for People:** Emphasized the importance of addressing employees' needs and well-being.
- **Team Management Style:** Combined high concern for both production and people to achieve optimal results and high satisfaction.

Their model reinforced the idea that effective leadership involves balancing task and people-oriented approaches. How you interact with your people has everything to do with outcomes. You can achieve results and treat people in a manner that keeps them wanting to work with and for you. Most of us can relate to what sacrificing one for the other looks like, and it's never pretty. Most first-time supervisors who become responsible for leading their former teammates errantly think they will get the best outcomes by continuing to be a great friend to everyone in the same way they always had. This undermines structure, accountability, and results for relationships, becoming a lesson that is usually learned the hard way by all. Blake and Mouton's approach underscores that balance is necessary.

Hersey and Blanchard: Situational Leadership

Hersey and Blanchard's Situational Leadership model, introduced in 1969, emphasized adapting leadership styles to the situation and the readiness of followers:

- **Task Readiness:** Recognized that people need more structure when new to a task.
- **Adaptability:** Suggested that leadership style should evolve based on followers' experience and competence.
- **Encouraging Independence:** Advocated for encouraging experienced followers to take ownership and act independently.

This model highlighted the importance of situational awareness in leadership, encouraging leaders to be flexible and responsive to their team's needs.

Needs Theories of Motivation: Maslow, Herzberg, Alderfer, and McClelland

The needs theories of motivation provide insights into what drives people to take action They advocate the importance of creating conditions that allow employees to satisfy their needs, leading to increased motivation and engagement:

- **Maslow's Hierarchy of Needs:** Identified a progression of needs from physiological to self-actualization.
- **Herzberg's Two-Factor Theory:** Distinguished between hygiene factors that prevent dissatisfaction and motivators that promote satisfaction.
- **Alderfer's ERG Theory:** Simplified Maslow's hierarchy into three core needs: existence, relatedness, and growth.
- **McClelland's Acquired Needs Theory:** Focused on the need for achievement, affiliation, and power.

The importance of leaders creating opportunities for their people to feel good about themselves, be recognized by others, and achieve on the job cannot be overstated.

Process Theories of Motivation

Process theories of motivation, such as expectancy theory and equity theory, emphasize the role of perceptions in motivation. They acknowledge that people's perceptions of fairness and expected outcomes significantly impact their motivation and behavior:

- **Expectancy Theory:** Suggests that people are motivated when they believe their efforts will lead to desired outcomes.
- **Equity Theory:** Focuses on the importance of fairness and perceived equity in motivation.

We've all heard it – perception is reality. Think of the worker who winds up feeling that even if they try, it won't matter anyway. As a leader, do you help this worker see things from a different viewpoint or let them slowly fail and go away? Your role is to influence and clarify when perceptions are inaccurate.

CONTEMPORARY IDEAS IN MANAGEMENT AND LEADERSHIP

Today's thought leaders are contributing to management and leadership best practices for modern organizations. These practices are being utilized to drive organizational growth and development, foster stronger organizational culture, and get the best with and through people.

Amy C. Edmondson: Psychological Safety

Professor Amy C. Edmondson, Harvard Business School, introduced the concept of psychological safety in the 1990s: Her focus is on two areas:

- **Psychological Safety:** Ensures that employees feel safe to express themselves without fear of punishment.
- **Open Communication:** Encourages transparency and learning from frontline workers.

Edmondson's work emphasizes the importance of creating a safe environment where employees can openly share ideas and concerns, leading to innovation and continuous improvement. Otherwise, how will you get open communication about imminent hazards if people fear reprisal?

Change Management Theories

Change management theories, from Kurt Lewin to John Kotter, emphasize the importance of communication in successful change efforts. Make no mistake, implementing a new technology, especially a wearable, involves successfully managing people through change. Regardless of the model or theory that may be your favorite, all are consistent in the importance of:

- **Clear Communication:** Ensures that employees understand the reasons for change.
- **Inclusive Approach:** Involves employees in the change process to gain their commitment and support.

These theories highlight that effective communication and inclusion are critical at each stage of the process to overcoming resistance and achieving lasting organizational change.

Deming and Total Quality Management (TQM)

While the number of fundamental principles from W. Edwards Deming's Total Quality Management (TQM) varies depending on where you look, five that are most essential for leading companies:

- **Continuous Improvement:** This is an essential act of management that aims to optimize the company's overall performance and bring about lasting positive change. Encourages ongoing efforts to improve processes and quality.

- **Employee Involvement:** All employees should participate equally in achieving mutual goals and consider themselves an integral part of the organization. Stresses the importance of including workers in decision-making.
- **Fact-based Decision-Making:** Making decisions based on data-based knowledge makes it easier to satisfy customers.
- **Leadership:** Leadership provides direction, motivation, and guidance to the organization's quality improvement efforts.
- **Communication:** Effective communication of information helps ensure that everyone is working toward the same goals and following the same standards.

TQM and Deming's contributions underscore the value of a culture of improvement and the vital role of communication and engagement in achieving excellence. We need to view safety management not as a series of checklists and functions to be performed to achieve compliance, but rather as an ongoing process of improvement that uses data evidence to make decisions.

Understanding Human Factors

Human factors refer to the study of how people interact with their environments, tools, and systems. These include:

- **Ergonomics:** Designing workplaces that fit the needs and capabilities of workers.
- **Cognitive Load:** Managing the amount of mental effort required by tasks.
- **User-Centered Design:** Creating systems and processes that prioritize the user's experience.

Understanding human factors helps leaders create environments that enhance productivity, safety, and well-being. Data evidence can support decisions about work design to ensure safety. An understanding of how people process information and make decisions certainly can inform better safety practices.

Systems Thinking

Systems thinking involves understanding the organization as a complex system of interconnected parts:

- **Holistic View:** Seeing the organization as a whole, rather than a collection of separate parts.
- **Interdependencies:** Recognizing how different components of the organization influence one another.
- **Feedback Loops:** Identifying and understanding the cycles of feedback that affect system behavior.

Systems thinking enables leaders to identify underlying patterns and root causes of issues, leading to more effective problem solving and strategic decision-making.

It is incumbent upon today's safety and operations leaders to consider how all parts of the organization may be affected by changes to systems because this impacts safety and productivity.

Mental models and understanding how our brains work

Mental models are deeply ingrained assumptions and generalizations that influence how we understand the world:

- **Cognitive Biases:** Recognizing biases that affect decision-making and behavior.
- **Neuroscience:** Understanding how brain functions impact learning, memory, and emotions.
- **Adaptive Thinking:** Developing the ability to adjust mental models in response to new information and changing circumstances.

Work on the front lines is never as it is imagined or planned because employees are constantly adapting and almost always are successful. They are real-time risk managers, constantly making decisions on how to balance priorities and achieve expected outcomes safely. Yet we give them rules, policies, protocols, and standards to follow and seemingly never stop piling on. By understanding mental models and brain functions, leaders can better appreciate how workers go about processing information and making decisions.

KEY TAKEAWAYS

- Over the past century, we have learned a great deal about leading and motivating effectively through understanding human nature and psychology.
- By incorporating insights from these early theories as well as contemporary ideas, leaders can create environments where employees are motivated, engaged, and empowered to contribute to the organization's success.
- The evolution of management thinking teaches us that while tools and technologies may change, the fundamental principles of effective leadership remain constant.

BIBLIOGRAPHY

"Employee Engagement." *Gallup*, 2024, https://www.gallup.com/workplace/349484/state-of-the-global-workplace.aspx

Part IV

The Future of Work, Tom West

18 Factories of the Future – Integrating the Connected Worker in Industry 4.0

The concept of the factory has undergone significant transformations over the past century, evolving from the simple assembly lines of the Industrial Revolution to today's highly automated and digitized production environments. With the advent of Industry 4.0, the factory is once again poised for a revolution – one that integrates advanced technologies like the Internet of Things (IoT), artificial intelligence, and wearable devices to create smarter, safer, and more productive industrial environments.

Wearable technology, such as we've developed at MākuSafe, can play a crucial role in this transformation. By enabling the connected worker to interact seamlessly with the automated and connected systems around them, these devices ensure that the human element remains integral to the factory of the future. Our clients have led us to explore how wearable tech can go beyond safety to enhance productivity and efficiency in Industry 4.0, highlighting the interdependence of these factors and the challenges that lie ahead.

THE PROMISE OF INDUSTRY 4.0: SAFER, ADAPTIVE, AND MORE PRODUCTIVE FACTORIES

Industry 4.0 represents a paradigm shift in manufacturing, where technology is not just an aid but an inherent aspect of the production process. The factory of the future is envisioned as a highly adaptive and interconnected environment, where machines, systems, and humans work in harmony to achieve unprecedented levels of efficiency and safety.

Wearable technology like MākuSafe is perfectly suited for this vision by integrating workers into the digital fabric of the factory. In our case, devices can monitor environmental conditions, detect potential hazards, and even adapt the surrounding systems to ensure a safer work environment. For instance, environmental sensors in wearables can detect when temperatures rise beyond safe levels and, when integrated into facility systems, automatically activate air handlers to cool the area, thereby preventing heat-related incidents.

But the potential of wearable tech goes beyond just safety. By collecting data on worker interactions with machines and systems, these devices can help optimize

DOI: 10.1201/9781003432302-22

workflows, reduce downtime, and increase overall productivity. In an Industry 4.0 environment, the factory is no longer a static entity but a dynamic system that continuously adapts to the needs of both workers and machines.

INTERDEPENDENCE OF SAFETY, EFFICIENCY, AND PRODUCTIVITY

In the traditional view of manufacturing, safety, efficiency, and productivity are often seen as separate, sometimes competing, priorities. However, Industry 4.0 challenges this notion by demonstrating the interdependence of these factors. When safety is enhanced through technology, efficiency and productivity naturally follow.

For example, consider a scenario where wearable devices are used to monitor the presence and activity of workers on a production line. If a worker enters a hazardous area, the system can automatically slow down or stop nearby machines to prevent accidents. This not only protects the worker but also ensures that the production process is not disrupted by incidents, thereby maintaining efficiency and productivity.

Moreover, data collected from wearables can provide insights into how work is being performed, allowing for more accurate cost accounting and quality control. By understanding which teams or workers are more productive and why, managers can make informed decisions to improve overall factory performance. This interdependence highlights the importance of viewing safety, efficiency, and productivity as mutually reinforcing goals in the context of Industry 4.0.

If our goal is continuous learning in order to improve systems and create resilience in processes, and we understand that no one knows the realities of work as well as the people performing it every day, then enabling easy communication and creating this new habit has such tremendous potential. Forward-thinking organizations recognize that this is so valuable that they no longer want to wait for shift change meetings to hear improvement ideas about safety, quality, and production, but instead want that input during a shift in real time so they can make rapid changes immediately.

CHALLENGES OF INDUSTRIAL DIGITAL TRANSFORMATION

While the potential benefits of Industry 4.0 are clear, the path to achieving this vision is fraught with challenges. One of the biggest hurdles is the digital transformation of industrial operations, particularly the integration of data from various sources into a centralized, actionable system.

Many organizations have made significant investments in Industrial IoT (IIoT) devices, collecting vast amounts of data from machines and systems. However, this data is often fragmented, incomplete, or not timely enough to be useful. In contrast, data from wearable devices like MākuSafe can provide real-time, contextual information about workers' interactions with their environment. This data is not only easier to collect and centralize but also offers insights that might be more challenging to obtain from traditional IIoT efforts.

For instance, while some organizations have spent years trying to prove value in use cases and connect all their IIoT devices, MākuSafe can typically be deployed in a large facility and start collecting data within half a day. The data can be exchanged with other cloud platforms. This rapid deployment and ease of use make wearable technology a compelling option for companies looking to enhance their digital transformation efforts without the complexity and delays associated with IIoT integration.

A RESPONSIVE CHOCOLATE FACTORY

A classic episode from the 1950s television show *I Love Lucy* provides a humorous yet insightful illustration of the challenges that can arise in a production environment. In the episode, Lucy and her friend Ethel work on a chocolate factory production line, wrapping candies as the conveyor belt speeds up uncontrollably. Unable to keep pace, the duo resorts to stuffing chocolates into their mouths and hats, leading to chaos on the production floor.

While this scenario is played for laughs, it highlights a very real issue that persists in many factories today – the inability of production systems to adapt to the presence and pace of human workers. In the factory of the future, such situations could be avoided through the use of wearable technology. Imagine a production line that can detect the number of workers present and automatically adjust its speed and volume to match their capacity, ensuring both safety and productivity.

Moreover, wearable devices could track which workers are assigned to which lines, how long they have been working, and what tasks they are performing. This data could be used to conduct efficiency studies, optimize labor allocation, and ensure that workers are compensated accurately based on the complexity of their tasks. The insights gained from wearable tech could help managers understand why certain teams are more productive than others and identify areas for improvement.

ENHANCING SAFETY AND PRODUCTIVITY WITH WEARABLE TECH

The ability of wearable devices to detect worker presence has significant implications for safety and productivity. For example, if a worker approaches a machine too closely, the wearable could trigger the machine to slow down or stop, preventing accidents. Additionally, by integrating with cloud-based systems like HR databases,

FIGURE 18.1 Chapter 18 sidebar QR chaos-to-control – Sidebar Video – From Chaos to Control: How Wearable Tech Can Transform Production Efficiency https://makusafe.com/blog/chaos-to-control/

wearables could verify whether a worker's credentials, training, or licenses are up to date before allowing them to operate certain equipment or enter secure areas.

Traffic mapping is another area where wearables can have a substantial impact. By detecting the movement patterns of workers throughout the factory, lean process engineers can identify wasted time or unnecessary travel, leading to more efficient workflows. This data-driven approach to process optimization is a key component of Industry 4.0, where continuous improvement is achieved through the intelligent use of technology.

The next frontier will be harnessing the power of AI fueled by all of this data being gathered from devices and workers. By utilizing generative and predictive modeling, AI can provide insights for companies about where and how they can optimize their operations.

As factories evolve, so too do the expectations of the workers who operate within them. Today's employees are increasingly tech-savvy and expect the tools they use to make their jobs easier, safer, and more connected. Wearable technology meets these expectations by providing leaders with real-time information about their environment, helping them keep workers free from hazards and enabling them to communicate with their supervisors.

WEARABLES AS HUMAN-CENTRIC TECHNOLOGY

The University of Tennessee Reliability & Maintainability Center (RMC) installed and tested MākuSafe technology in their training factory for College of Engineering students as well as industry and government professionals. The implementation was overseen by Dr. Klaus M. Blache, RMC Director and Research Professor.

Dr. Blache has written extensively about various areas of manufacturing and continuous improvement for both processes and people. He is a champion of safety and change management, having worked for General Motors and served as chairman of the Society of Maintenance & Reliability Professionals. In his view, technology like MākuSafe is a significant contributor to workplace safety.

"The holistic well-being of employees will continue to play a greater role in employee safety, engagement, and productivity. A key attribute of Industry 4.0 is being more 'human-centric'. That refers to using digital technology to enhance and enable people to work at their best capability and contribute to human ingenuity."

As a seasoned professional who has authored two books, lectured globally, and helped numerous Fortune 500 companies with strategies and implementation, Klaus recognizes how wearables will help transition the current workforce skills and culture to Industry 4.0 and beyond.

"Based on my over 30 years of working across manufacturing, ergonomics/safety, maintenance & reliability, and now in academia helping industry and government agencies improve their operations, I view MākuSafe as a unique value-added choice for improving health, safety and productivity. MākuSafe data provides valuable real-time EHS insights in potential areas of concern to improve worker safety, absenteeism, turnover and overall well-being."

CONCLUSION

In the factory of the future, wearable tech is not just a tool for management but an empowering device for workers. By integrating the connected worker into the automated and digitized factory environment, wearable devices like MākuSafe help bridge the gap between human labor and machine efficiency. This symbiosis ensures that the factory of the future is not only more productive but also a safer and more satisfying place to work. Wearable tech can play a crucial role in creating a factory environment that is adaptive, responsive, and inherently safe – a true reflection of the promise of Industry 4.0.

KEY TAKEAWAYS

- The integration of wearable technology into Industry 4.0 represents a significant leap forward in the evolution of the factory.
- By connecting workers to the digital systems around them, these devices enhance safety, productivity, and efficiency.
- Achieving this vision requires overcoming challenges related to data integration and digital transformation.

19 The Future of EHS Embraces Technology and New Management Approaches

In the ever-evolving landscape of workplace safety, the future of Environment, Health, and Safety (EHS) management is upon us. It's time to shed outdated practices and embrace innovation to ensure the well-being of workers while enhancing operational efficiency. Our parting chapter explores the pivotal role of technology and new management approaches in shaping the future of EHS, highlighting the imperative for organizations to adapt or risk falling behind.

EMBRACING INNOVATION IN EHS

Historically, safety management has operated in isolation, disconnected from the technological advancements transforming other sectors. Today, the convergence of cutting-edge technology and progressive management philosophies is driving a paradigm shift in EHS practices. Organizations must recognize the interdependent nature of safety, productivity, and efficiency. It is crucial to move away from siloed objectives toward holistic approaches that prioritize worker well-being without compromising operational excellence.

Innovative solutions are now available that integrate safety management with broader operational goals. These solutions include sophisticated data analytics, real-time monitoring systems, and AI-driven insights that help organizations anticipate and mitigate risks before they materialize. This evolution is not merely about adopting new tools but about fostering a culture that values and applies technological advancements to continuous improvement.

HARNESSING THE POWER OF DATA

One of the most significant developments in modern EHS is the ability to gather, analyze, and utilize data to optimize operations and ensure worker safety. Imagine an advanced manufacturing facility where systems and equipment adapt in real time to the workers interacting with them, creating a symbiotic relationship between safety and productivity.

Real-time data collection and analysis empower organizations to identify patterns, predict potential hazards, and implement preventative measures. For instance, wearable technology can monitor individualized heat indexes, sound exposure, air quality,

DOI: 10.1201/9781003432302-23

and physical strain. This data-driven approach not only enhances safety but also improves efficiency and productivity by minimizing downtime and preventing accidents before they happen.

However, the integration of technology in EHS is not without its challenges. Common concerns regarding technology persist, including worker receptiveness, privacy issues, and potential impacts on organizational culture and morale. Addressing these concerns requires transparent communication and a commitment to using data responsibly and ethically.

HUMAN AND ORGANIZATIONAL PERFORMANCE

Central to the evolution of EHS is a shift toward human-centric approaches that focus on understanding the realities of work as it is performed. Safety luminaries such as Todd Conklin, Sydney Dekker, and Bob Edwards emphasize the importance of moving beyond blame and punishment toward cultivating a culture of learning and resilience. Safety is not merely the absence of incidents, but the presence of capacity in redesigned systems for people to fail safely, even on imperfect days.

Frameworks like Safety 2.0 and similar concepts underscore the need for organizations to build resilience and capacity into work processes, enabling workers to adapt and respond to unexpected challenges. This involves creating an environment where workers feel empowered to report hazards, share insights, and participate actively in safety initiatives. The goal is to learn from everyday work experiences and continuously improve safety practices with high engagement and participation.

DATA-DRIVEN DECISION-MAKING

The advent of wearable technology and advanced analytics empowers organizations to gain real-time awareness of risks and potential hazards. By collecting data associated with leading factors, organizations can proactively implement controls and improve mitigation efforts.

For example, at MākuSafe, our workforce wearable tech gathers data in real time about potential hazards. This includes individualized heat index, sound exposure, air quality, physical work effort, and ergonomic risks. Our clients routinely report significant improvements in safety metrics. Many have seen reductions in incident and claims frequency and severity exceeding 50–90%, or strain and exertion incidents dropping to zero within their first 12 months of use.

Data-enabled decision-making not only increases the effectiveness of management processes but also closes the continuous improvement loop, ensuring progress toward desired outcomes. By using real-time data, organizations can implement targeted interventions, monitor their effectiveness, and adjust strategies as needed. This iterative process supports a proactive safety culture where risks are managed dynamically and improvements are sustained over time.

ELEVATING FRONTLINE VOICES

Effective EHS management goes beyond compliance, amplifying the voices of workers. Technology can serve as a tool for empowerment, enabling employees to actively participate in safety initiatives and contribute to a culture of continuous improvement. Moreover, initiatives aimed at enhancing ESG (environmental, social, and governance) measures can bolster organizational performance, win new business, and attract and retain top talent in an increasingly competitive landscape.

By giving workers a platform to share their experiences and insights, organizations can gain a deeper understanding of the realities of work. This inclusive approach not only improves safety outcomes but also enhances worker engagement and morale. When workers feel heard and valued, they are more likely to buy into safety initiatives and contribute to a positive safety culture.

Furthermore, the integration of technology in EHS can support broader organizational goals, such as sustainability and social responsibility. By aligning safety initiatives with ESG objectives, organizations can demonstrate their commitment to ethical and responsible business practices, which can enhance their reputation and competitiveness in the market.

TOWARD STRATEGIC TRANSFORMATION

Ultimately, the goal of modern EHS management is to facilitate organizational transformation and sustainable competitiveness. By embracing technology and adopting progressive management approaches, organizations can move beyond traditional metrics of success toward achieving strategic objectives. From reducing incident rates to enhancing operational efficiency and productivity, the integration of technology and new management paradigms holds the key to a safer, stronger, and more resilient future.

Organizations that prioritize EHS as a strategic imperative can also achieve outcomes that extend beyond compliance. By integrating safety into their core business processes and applying technology to drive continuous improvement, they can create a culture of excellence that supports long-term success. This requires a commitment to innovation, collaboration, and a willingness to adapt to changing circumstances.

In conclusion, the future of EHS management is characterized by innovation, collaboration, and a commitment to continuous improvement. Organizations must embrace technology, leverage data-driven insights, and prioritize human well-being to thrive in an increasingly complex and dynamic world. By doing so, they can not only ensure the safety and well-being of their workforce but also position themselves for long-term success in a rapidly evolving business landscape.

As we move forward, let us remember that the journey toward a safer and more sustainable future is ongoing. It is through collective efforts and a shared commitment to excellence that we can truly realize the full potential of EHS in the 21st century. The future of EHS is not just about preventing incidents but about creating an environment where safety, productivity, and efficiency are mutually reinforcing. By embracing innovation and adopting a holistic approach to safety management, we can build resilient organizations that are equipped to navigate the challenges and opportunities of the future.

Part V

Appendix 1
Case Studies

In this appendix, we provide an overview of several case studies conducted recently using our wearable technology and have summarized key findings. These reports were all published by third parties or the users themselves. Links are included to access the full report for each.

The latest MākuSafe case studies, whitepapers, and reports will continue to be updated regularly and can be accessed here along with relevant content that may be of interest.

FIGURE A.1 QR Code 1 to https://makusafe.com/downloads/

NATIONAL SAFETY COUNCIL CASE STUDY OVERVIEW: NFI INDUSTRIES' USE OF MĀKUSAFE WEARABLE TECHNOLOGY

Introduction The National Safety Council's Work to Zero initiative, in partnership with SafetyTech Accelerator, conducted a case study highlighting NFI Industries' innovative approach to reducing musculoskeletal disorders (MSDs) in the workplace. MSDs are a significant concern, impacting nearly one quarter of the global workforce, with injuries such as strains, carpal tunnel syndrome, and tendonitis resulting from repetitive movements, awkward postures, and forceful exertions.

The Solution: MākuSafe Wearable Technology To address these challenges, NFI Industries piloted the MākuSafe system, a safety data and analytics solution. MākuSafe uses wearable technology to monitor and analyze worker movements, providing real-time insights into factors that can contribute to MSDs. The system collects data on physical exertion, movement types, incident timing, and location, which safety leaders use to better understand potential risks and support employees in preventing injury.

Pilot Program and Results NFI deployed 120 MākuSafe wearables across three locations, focusing on material-handling workers engaged in physically demanding tasks like loading and unloading. By combining the data provided by MākuSafe with targeted lifting training and behavior-based safety strategies, NFI achieved a remarkable 74% reduction in injuries at the pilot sites. Additionally, employee surveys showed a high level of favorability, with workers voluntarily embracing the wearables as part of the company's safety culture.

Key Lessons Learned: As NFI continues to expand the use of MākuSafe, several lessons emerged:

- Integrating behavior-based safety processes can enhance the organization's safety culture.
- Transparent communication is essential – employees should understand the technology's purpose and how data will be used.
- Regular feedback from both employees and managers is critical for continuous improvement.
- IT department engagement is crucial for addressing configuration, security, and compatibility concerns.
- It's important to fully understand the technology's capabilities and limitations to avoid being misled by unnecessary features.

Conclusion NFI's adoption of MākuSafe technology showcases the potential of wearable devices and data-driven insights to significantly reduce workplace injuries, creating a safer and more engaged workforce.

About NFI NFI Industries is a third-party logistics provider headquartered in Camden, New Jersey, employing over 16,800 associates nationwide. Their supply chain solutions span domestic and international operations.

For a deeper dive into this case study, download the full report here

FIGURE A.2 QR Code 2 to https://makusafe.com/wp-content/uploads/2023/06/WTZ-NSC-MakuSafe-Case-Study.pdf

The National Council on Compensation Insurance (NCCI) published research in 2024 exploring the impact of safety technologies on workplace environments, including wearables, AI, and computer vision systems. The report, part of a series, emphasized the rapid evolution of workplace safety and the importance of technological innovations in workers' compensation. NCCI conducted interviews with various stakeholders, including employers using innovative safety technology, to understand the implementation challenges and the benefits these technologies provide.

Key takeaways from the research include:

- Safety technology requires a strong partnership between the employer, technology provider, and insurer.
- The cost of implementing such technologies is not always a barrier, as the long-term savings in workers' compensation and workplace safety outweigh initial expenses.
- Employee buy-in is essential, achieved through transparency, education, and company leadership adopting the technology themselves.
- The technologies have resulted in both improved safety and increased productivity in industries like manufacturing and warehousing.

One prominent example highlighted was Life Line Emergency Vehicles, a company utilizing MākuSafe wearables. Initially, Life Line started with 40 devices, which expanded to 160 after seeing positive outcomes. Key successes include:

- **Data-Driven Insights:** Life Line began conducting daily reviews of wearable data, which led to monthly meetings with their technology provider and quarterly insurer reviews. These discussions identified ergonomic challenges, resulting in solutions like the installation of lift tables and changes to work processes.
- **"Good Catches":** Over 335 potential hazards, or "good catches," were reported, with 82% successfully addressed. This proactive hazard identification system prevented injuries and helped maintain a safer work environment.
- **Significant Cost Savings:** Workers' compensation costs were reduced by 60% between 2021 and 2023, highlighting the financial advantages of wearables.
- **Improved Workplace Safety:** Life Line saw fewer lost time hours, reduced trip hazards, and cleaner, more efficient workspaces.
- **Employee Engagement:** By promoting transparency and involving management in wearing the devices, Life Line addressed privacy concerns and fostered employee buy-in, making the wearables a standard part of their safety culture.

Overall, the Life Line experience underscores the transformative role that wearable technology can play in enhancing workplace safety, reducing costs, and improving operational efficiency.

For a deeper dive into this case study, download the full report here

FIGURE A.3 QR Code 3 to https://makusafe.com/blog/enhancing-workplace-safety-with-wearable-technology-a-case-study/

Workers Compensation Insurance Carrier Findings with Wearable Safety Technology Solution

A large worker's compensation insurance carrier piloted the MākuSafe safety technology with six of its policyholders having industrial operations over the course of two to six months, resulting in significant real reductions in claims. This findings report highlights some of the hazards identified and the overall results achieved during the test periods, including claims frequency reduced by 50% and claims severity reduced by 90%.

Company A (@ 6 months)

- 87% decrease of indicators per active device from first half (1,206) to second half (159) of the trial
- 90% reduction in work comp claim severity – $3,089before MākuSafe, $311 after
- Employee voice memo alerted management of equipment overload switch tripping, which likely prevented a $10,000 (average for their SIC) equipment breakdown claim or possible fire.

Company B (@ 6 months)

- 30% decrease of indicators per active device from first half (454) to second half (317) of the trial
- 30% reduction in work comp claim severity – $120,575 before MākuSafe and $84,091 after
- Indicator disparity of work cells conducting the same work gave management the insight to investigate and identify ergonomic hazards present in the high indicator cell. If the corrective action prevents just a single overexertion injury in the work cell, that would be an$18,000 loss avoided (the company's average overexertion).

Company C (@ 3 months)

- 39% decrease of indicators per active device from first half (53.2) to second half (32.5) of the trial
- 91% reduction in work comp claim severity – $10,051 before MākuSafe and $899 after

- Employee voice memos were used to report slip-fall hazards and other employees not wearing the required slip-resistant shoes. A single slip-fall injury averages $9,630 for this manufacturer.

Company D (@ 2.5 months)

- 3% decrease of indicators per active device from first half (100) to second half (96) of the trial
- 70% reduction in work comp claim severity – $24,320 before MākuSafe and $7,317 after
- Number of indicators leading management to completely overhaul their ergonomics program and collaborating with workers on task analysis to reduce unnecessary motions and force, which may result in significant operational improvements and lower workcomp mod (currently 1.16)
- High CO_2 levels help identify scrap wood burners not venting properly, where correcting chimney efficiency will have a positive impact on worker health and performance.

Company E (@ 2 months)

- 46% increase of indicators per active device from first half (57.5) to second half (84.2) of the trial
- 0 work comp claims since MākuSafe, but no claims six months prior either; they do average $9,000/year
- A device detected a trip. Management investigated and found the worker tripped on an air hose laying on the ground, so a hose reel was installed to eliminate the hazard. Preventing just a single trip-fall due to this improvement would be a savings of $10,000 (the average trip-fall for their SIC group).

Company F (@ 2 months)

- 8% decrease of indicators per active device from first half (27.6) to second half (25.5) of the trial
- Work comp claim analysis not available, not a policyholder

For a deeper dive into this case study, download the full report here

FIGURE A.4 QR Code 4 to https://makusafe.com/wp-content/uploads/2023/03/Work-Comp-Carrier-Findings-Program-Summary-1.pdf

CASE STUDY BY WORKERS COMPENSATION INSURER: THIRD PARTY LOGISTICS

Overview: Wearable Technology Case Study with Conner Strong & Buckelew

This case study highlights a successful pilot program of MākuSafe wearable technology, led by Conner Strong & Buckelew, in collaboration with a leading third-party logistics provider. The aim was to enhance workplace safety by addressing known risks, such as lifting-related injuries, and identifying additional hazards like slips, repetitive motion, and heat exposure.

Key Findings:

- The logistics provider sought improved visibility into safety risks, beyond lifting and pushing, to enhance their employee safety protocols.
- Conner Strong & Buckelew recommended MākuSafe as a wearable technology solution to monitor workplace hazards, providing real-time data on environmental and ergonomic risks.
- By combining wearable technology with a lifting mechanics training program, the company improved employee behavior and training, resulting in a **56% reduction in lifting-related injuries**.
- The company saved **$450,000 in Workers' Compensation losses** in just one year.
- The pilot data also reinforced the company's behavior-based safety program, offering actionable insights into employee safety practices.

For a deeper dive into this case study, download the full report here

FIGURE A.5 QR Code 5 to https://makusafe.com/news/case-study-conner-strong-buckelew-finds-incredible-results-with-wearable-tech/

Case Study Summary: Data-Driven Worker Safety Improvement at Tom's Quality Millwork (TQM)

Background: This case study highlights the use of MākuSafe wearable technology at Tom's Quality Millwork, a major hardwood manufacturer, to improve workplace ergonomics and reduce worker injuries. A team of undergraduate industrial engineering students from the University of Wisconsin-Madison collaborated with TQM to analyze data collected from MākuSafe devices worn by shop floor workers. The aim was to reduce risks associated with high-force tasks like pushing, pulling, and repetitive movements, which are linked to long-term injuries and worker compensation claims.

Key Findings: The study focused on TQM's Line 3, which operates on two shifts. Workers rotate every 2.75 hours across four workstations, each handling tasks of varying difficulty depending on the product size. The MākuSafe data identified patterns in high-risk activities, including:

1. **Push/Pull Activities:** The data revealed frequent and elevated occurrences of push/pull movements on certain days, correlating with the production of larger and heavier boards.
2. **Time and Position Impact:** The frequency of push/pull movements decreased throughout the day, and positions 2 and 4 were found to have significantly higher occurrences of these activities. Certain workers were also identified as experiencing more push/pull hits, suggesting a need for further ergonomics adjustments.
3. **Job Rotation and Fatigue:** The analysis demonstrated that task frequency and job rotation influenced the strain on workers. Push/pull hits were notably higher during the morning rotations compared to the afternoon, indicating worker fatigue as a contributing factor to safety risks.

Recommendations: The student research team developed several practical recommendations to improve worker safety and reduce injury risk:

1. Introduce anti-fatigue mats at key positions.
2. Adjust the placement of scrap bins to reduce unnecessary movements.
3. Shorten the shift rotation length from 2.75 hours to 2 hours.
4. Decrease task frequency, especially for repetitive high-force tasks.
5. Ensure exclusive use of counterbalance carts for final product sorting to reduce manual handling.
6. Reorganize stacking carts to minimize strain.
7. Limit the number of boards handled simultaneously.

Outcomes: By following these data-driven and ergonomically informed recommendations, TQM is expected to lower the risk of worker injuries, reduce worker compensation claims, and maintain productivity on the shop floor. The implementation of these safety improvements offers a clear example of how wearable technology can drive meaningful workplace changes.

For a deeper dive into this case study, download the full report here

FIGURE A.6 QR Code 6 to https://makusafe.com/news/data-driven-worker-safety-improvement-at-toms-quality-millwork/

Construction General Contractor Safety Success with MākuSafe Wearable Technology

MākuSafe collaborated with the forward-thinking leadership team at The Weitz Company, on their Mission Critical projects. Through this partnership, Weitz achieved a remarkable milestone: a 0.0 Total Recordable Incident Rate (TRIR) over more than 31 months on sites where MākuSafe was used. This achievement is a testament to their dedication to worker safety, their embrace of innovative solutions, and their commitment to operationalizing technology beyond the pilot stage.

Weitz's Approach to Safety with MākuSafe The Weitz Company has a long-standing commitment to safety, and as their focus on Mission Critical and data center projects grew, they sought out new technologies to protect their workforce. MākuSafe's wearable devices collect and analyze data to help safety teams predict potential hazards before incidents occur. In a large multi-employer construction environment with many moving parts, small near-miss incidents often go unreported. MākuSafe's technology helps prevent these unreported near-misses from escalating into serious accidents by providing real-time data on conditions like environmental factors, motion, and worker location.

In a practical example shared by a Senior Project Manager, MākuSafe data revealed multiple slipping motions near a site entrance during a snowy day. Thanks to the alerts, Weitz's logistics team quickly addressed the hazard, preventing potential injuries before they could occur.

Meeting ESG Expectations The company's innovative approach to safety also aligns with the growing environmental, social, and governance (ESG) expectations from their clients. As more companies seek to ensure that their partners prioritize worker well-being and sustainable practices, proactive use of MākuSafe technology supports these objectives. By utilizing real-time safety data and implementing preventative measures, Weitz demonstrates their commitment not only to safety but also to the broader social responsibility of protecting workers and minimizing incidents – a key ESG requirement.

Key Benefits: Safety, Efficiency, and Health Weitz has deployed MākuSafe across multiple job sites, including their trade partners and subcontractors. This technology doesn't just protect workers from physical hazards but was also instrumental during the COVID-19 pandemic. With contact tracing capabilities and worker density mapping, MākuSafe provided critical data to reduce the risk of viral transmission on job sites, helping maintain health and productivity.

A Culture of Safety The Weitz leadership emphasizes the importance of embedding safety into the company's culture. By using MākuSafe technology, they've successfully engaged workers, improved safety performance, EMR and TRIR have improved significantly. As the first construction company to adopt MākuSafe, Weitz has continued to expand its use across various business units, setting an industry standard for safety.

The partnership between MākuSafe and The Weitz Company demonstrates the power of innovation in creating safer, more efficient work environments while also meeting the ESG expectations of clients and stakeholders.

For a deeper dive into this case study, download the full report here

FIGURE A.7 QR Code 7 to https://makusafe.com/news/makusafe.contributes-to-0-0-trir-on-data-center-construction-projects-the-weitz-company/

Appendix 2
Wearable Technology Endorsements from Industry Professionals

We are grateful that so many of our peers were willing to provide their thoughts on how MākuSafe is influencing the safety landscape globally. We were so overwhelmed by responses from our contacts and mentors in the industry and related fields that we wanted to include them, in their entirety, in this book.

Along the way, we have learned from each and every one of these folks, and our approach has been shaped by them. Our hope is that you take something from their comments here. We hope that you'll find it encouraging that so many thought leaders from the safety space and its fringes are seeing that safety has the opportunity for significant progress by embracing the principles and approaches we've suggested in this writing.

Gabe, Mark, and Tom

In today's rapidly evolving industrial landscape, safety management is undergoing a transformative shift – and *Safety and Wearable Technology* is the definitive guide to understanding how wearable technology is revolutionizing the field. As industries face growing pressures to improve safety outcomes, this book is a timely exploration of how data-driven insights from wearable devices can provide a new layer of protection for workers and companies alike.

This book, authored by experts Gabriel Glynn, Mark Frederick, and Tom West, is an indispensable resource for safety professionals, industry leaders, and anyone looking to understand the future of safety management. It breaks down complex topics like sophisticated sensors, IoT applications, and machine learning in a way that's accessible and actionable, making it clear why wearable safety technology is not just a trend – but a critical tool in reducing life-altering incidents and preventing fatalities.

Readers will appreciate how the authors go beyond technology itself to address the cultural and managerial shifts necessary to implement these innovations successfully. With chapters on green flags in safety culture, the imperative of preventing serious injuries, and the evolving role of safety leadership, the book provides a comprehensive roadmap for integrating wearable technology in ways that truly enhance safety outcomes.

In a world where predictive analytics and connected workers are becoming the norm, *Safety and Wearable Technology* underscores why embracing these advancements is essential – not only for compliance but also for creating safer, smarter, and more efficient workplaces.

Corrie Pitzer
CEO & Founder, Safemap International Pty. Ltd.

I worked for the Occupational Safety and Health Administration for many years, and it was our job to investigate workplace incidents and place blame on the employers. Yes, the employer was wrong (and cited) unless they could prove that the blame belonged to the employee! Todd Conklin convinced me, although after my retirement from OSHA, that "blame fixes nothing," "learning and improving is vital," and "learning is deliberate." But learning requires understanding and understanding requires data! Gabe, Mark, Tom and the entire MākuSafe team are committed to helping employers learn and improve through the deliberate acquisition and analysis of data. What a concept! As a 35+ year safety professional, I have recently seen the future and the future is MākuSafe!

Doug Fletcher, former OSHA Official
President, Fletcher Safety Consulting, Inc.
Host "What's The Hazard" Podcast

I highly recommend this book to anyone interested in the future of workplace safety and operational excellence. The authors brilliantly integrate Safety 2.0 concepts from thought leaders like Todd Conklin and Sidney Dekker with their own expertise in cutting-edge Industry 4.0 solutions. This book clearly demonstrates how these complementary systems can revolutionize safety management and business operations.

Chad Veach, CSP, ARM, CPCU
Assistant Vice President-Risk Improvement Innovation, EMC Insurance

Historically, safety tech has lagged behind, but this lack of "technical debt" offers a unique advantage. By embracing wearable technologies and predictive analytics, as outlined in *Safety and Wearable Technology*, safety leaders can bypass outdated systems and move directly into the future of proactive safety management. Claire Jackson's journey is a perfect example of how we can seize this moment to make workplaces safer. It's an essential read for anyone ready to embrace the next era of safety.

Scott Cuthbert
CEO & Co-Founder, Safeopedia Inc.
Canada

With over 35 years of experience leading safety and health programs, I've seen firsthand the complex challenges that safety practitioners face. As injury rates plateau and fatalities rise, professionals around the world find themselves at a critical juncture, asking: How can we break through the barriers to achieve the safety performance we desire?

Safety and Wearable Technology: Impact, Applications, and Implementation in Industry reveals the groundbreaking potential of wearable technology to predict and prevent incidents before they occur. By integrating real-time data, AI, and human insight, this book presents a proactive approach that's reshaping the future of safety. Technology is poised to take us closer to our goal of ensuring every worker returns home safely. Within these pages is a beacon of hope in reaching this lofty endeavor. Are you ready to embrace its potential? Your journey starts here.

Scott Gaddis
Vice President, Global Practice Leader, Safety and Health
Intelex Technologies

I've been tracking wearable tech for over a decade, starting from my involvement in an "emerging issues" discussion group at an insurance company. Since then, I've kept a close eye on the evolution of safety technologies and have spoken at numerous conferences on the subject. This book is an important addition to the conversation, providing clear insights into how wearable safety technology is changing the way we think about risk management.

Insurance entities are not just in the business of paying claims – they are in the business of protection. The ultimate goal is to predict and prevent losses, and wearable safety tech plays a critical role by providing the data we need to identify risks before they result in injuries. After all, you can't solve a problem you haven't identified and analyzed. This book equips readers with the tools to do exactly that.

Today's safety directors are more data-driven than ever, and wearable safety tech offers unprecedented insights. This tech allows us to passively gather metrics that would have taken days of observation in the past. By analyzing data on where risks occur, how often, and under what conditions, we can take action more quickly and effectively.

While this book will certainly be a go-to resource for safety professionals, it also offers significant value to anyone in operations. As I often say, safety, quality, and productivity are like a three-legged stool – when one improves, so do the others. The authors understand this, and their work reflects a broader perspective that touches on all aspects of business performance.

One of the key strengths of this book is its balanced approach. The authors not only explain the benefits of wearable tech but also address potential challenges and objections. For instance, I had always wondered why haptic feedback was left out of the MākuSafe device, but after reading the book, I understand the reasoning behind that decision. It forces readers to think more broadly and critically about wearables and how they can be best implemented in their own workplaces.

I first learned about the MākuSafe concept more than eight years ago, when it was still a sketch on a napkin. I knew it was a groundbreaking idea then, and this book reinforces that belief. The authors are among the smartest people I know, and every conversation with them leaves me more informed and more optimistic about solving workplace safety challenges. I highly recommend this book to anyone who wants to stay ahead of the curve in safety and risk management.

Chris Murphy, CSP, CRM, CPCU
VP Commercial Risk & Safety, Holmes Murphy

I have watched the MākuSafe technology platform grow and expand for more than four years. Every time I meet with the team, I marvel at the advancement in the technology and how they have really improved the platform over time. I am delighted that the company has made such a huge commitment to research and development, with nearly half of the workforce dedicated solely to product development. I think that is an amazing trait of a startup to continue to heavily invest in the product and not be lured solely by sales and marketing of the product.

Their technology platform has great architecture and can interface with many "dashboard" platforms to integrate with existing systems versus a standalone system. This equates to a better interface and higher likelihood of acceptance and sustainability in an organization.

Wearables are truly the future of health and safety. They provide the most proactive means to address hazards and exposures. Although near-miss reporting is a significant strength of my company, it requires a human act to complete, with many competing priorities. Wearables act independently and without prompting to provide nearly real-time evaluation of hazards. It is clearly the most proactive thing that can be done in the spectrum of injury prevention.

MākuSafe heavily invests in their customers as well, ensuring the product implementation and support teams are engaged in the launch and continued follow-up with sites. We have a site that is using the MākuSafe technology. Every aspect of the sales, installation, start-up, and user experience has been exceptional. The staff is top notch and really helpful. They are truly invested in the success of the product to work and make the user experience successful. Let's face it – workers' acceptance of new technology is mixed. For some, there are reservations on what is being monitored and how the data is being used. MākuSafe is very helpful at promoting worker acceptance through education, such that the worker embraces the use of the technology, knowing the nobler cause is their personal well-being/health and safety.

We have had some interesting revelations from our implementation and use of the MākuSafe. We had an operation that was seeing increased levels of VOCs in an area where no solvents are present. Through the proactive technology of MākuSafe, we were able to determine the source as workers using an air wand to remove some residual solvent from a degreasing operation. The air wand created the vapors that were detected by the wearable. Without this, we would have never known this exposure was occurring. Due to this proactive approach, we were able to look at the degreasing operation to determine options to reduce the residual solvent and eliminate the exposure to our employees.

In another operation, the wearable noted that one employee had higher energy expenditure and higher scores relative to ergonomic exposure. After reviewing the job, we noted the job was not set up well for workstation adjustability. The stature of the worker was not well suited to the height requirements since one of the job elements also included a hard-to-reach area. The local ergonomics team was able to review this and address the issue.

I think MākuSafe has become the market leader in wearable health and safety technology. Their platforms and user interface are excellent and only further evolve to become even better. Their commitment starts at the top, with the senior leadership of Gabe, Mark, and Tom. They are truly authentic in their leadership style and have built an excellent product and team. They are trustworthy and committed to excellence.

Mark Jones, PhD, MBA, CIH, CSP
EHS Director, Plastipak Packaging
Named one of America's Safest Companies by EHS Today
Past President and Senior Board Member, Michigan Safety Conference
30 years in Safety with Ford Motor Company

As a seasoned safety professional with over 40 years of experience, I have witnessed significant changes in the field. From my early days of using basic tools to measure flow rates, I now find myself immersed in cutting-edge concepts like the Fourth Industrial Revolution. While the methods may have evolved, the core purpose of safety remains unchanged – identifying and addressing the underlying causes of accidents.

Regardless of the specific exposure, be it chemical or physical, safety professionals strive to identify and mitigate the events at the bottom of the accident triangle. These events encompass the activities and actions that put employees at a higher risk of accidents. For instance, in the case of a chemical exposure, it could be the frequency of slips while carrying chemicals. By analyzing data, we can determine that employees carry chemicals 30 feet four times a day, resulting in 40,000 instances per year. Of these, there are 80 near-misses, 22 instances requiring medical attention, and 5 cases of missed work. As safety professionals, our goal is to reduce both the frequency of carrying chemicals and the likelihood of slipping with each occurrence.

Similarly, we address physical hazards, such as back injuries, by focusing on the frequency of lifting boxes. For example, individuals on a packaging line lift boxes 10 times per hour, resulting in 160,000 lifts annually. By examining the data, we uncover 140 near-misses, 18 instances requiring medical treatment, 9 cases of lost work days, and, unfortunately, 1 permanent disability. Our aim is to reduce the number of lifts and minimize actions that increase the risk of injury, such as twisting, lifting below the knees, and lifting above the shoulders.

Focusing on the bottom of the accident triangle has proven highly effective in reducing accidents and injuries. By addressing unsafe acts and conditions, we can prevent near misses, medically required injuries, time loss, and even fatal or permanent total disability incidents. The challenge lies in quantifying these elements accurately to determine where improvements should be made. Wearable technology presents a viable solution. For instance, MākuSafe's matchbox-sized wearable sensor platform allows us to collect data that helps identify problem areas and prioritize corrective actions. Unlike traditional safety methods that rely on observations, wearable technology provides reliable data to analyze critical behaviors, leading to informed decision-making. *Safety and Wearable Technology: Impact, Applications, and Implementation in Industry*, written by Gabriel Glynn, Mark Frederick, and Tom West, helps readers understand the importance and usefulness of wearable safety tech. It offers valuable insights for implementing this technology successfully in various industries.

This book helps bridge the gap between professionals who seek data-driven solutions and the potential of wearable technology. Divided into five parts, it covers the necessity of data in safety practices, explores the functionalities of wearables and machine learning, provides implementation insights, offers a glimpse into the future, and presents case studies demonstrating the diverse applications of wearable technology. The authors' credibility is unquestionable, with ample experience in implementing wearable technology solutions across various industries. By sharing their experiences with early adopters, they guide readers through successful implementation, helping them avoid costly mistakes.

However, the book goes beyond mere technological explanations. Rather than relying solely on beeping alarms to notify employees of wrongdoings, it focuses on utilizing wearable technology to eliminate exposure within the workplace. By targeting the unsafe actions and conditions that drive loss, this book equips safety professionals with the knowledge they need to reduce the likelihood of accident consequences.

Whether you are a safety professional seeking to enhance your understanding of wearable safety tech or an industry leader in need of practical implementation strategies, *Safety and Wearable Technology: Impact, Applications, and Implementation in Industry* is an essential resource. It empowers readers to comprehend the pros and cons of wearable safety tech and provides invaluable guidance on successfully integrating these technologies into their facilities. Prepare to embark on an enlightening journey that will revolutionize your approach to workplace safety.

Rick Fineman
Vice President of Loss Control
Berkshire Hathaway Homestate Companies
Workers Compensation Division

The secret to innovation isn't ideation – it's inspiration and iteration. The passionate team behind MākuSafe are pioneers behind the design and daily use of wearable technology that truly achieves their mission of making workplaces safer. The key to their success has been tight collaboration with a wide range of partners and the ability to translate learnings into product design over a number of years. I'm thrilled that all this hard-earned knowledge has been condensed and summarized into a highly readable book that will broaden the audience who may benefit from the experience and insights of the MākuSafe team. Highly recommend!

Rob Galbraith, CPCU, CLU, ChFC
Founder and CEO, Forestview Insights

Author of the bestseller *The End Of Insurance As We Know It: How Millennials, Insurtech, and Venture Capital Will Disrupt the Ecosystem*, Rob has been called "a global insurtech thought leader," "a mentor to many in the insurance industry," "the most interesting man in insurance," "the prophet," and "the WD40 of insurance."

As a 33-year safety professional who has been working with MākuSafe as a customer for seven years, I wholeheartedly endorse MākuSafe, a pioneer industrial safety technology. In an era where workplace safety is paramount, this startup stands out for its innovative approach to safeguarding employees and streamlining safety protocols.

MākuSafe has developed a cutting-edge solution that sets a new benchmark for safety technology. Their product leverages advanced sensors, real-time data analytics, and intuitive user interfaces to deliver unprecedented levels of employee protection and operational efficiency. With the ability to integrate these technologies into existing operational systems, they make it easier for organizations to proactively manage risks and ensure a safer working environment.

What truly sets MākuSafe apart is their commitment to both technological excellence and user-centric design. Their team of experts collaborates closely with industry leaders and customers to tailor solutions that address real-world challenges, ensuring their technology is practical and effective. This dedication to understanding and responding to the needs of their clients is a testament to their passion for enhancing workplace safety.

Using MākuSafe's safety technology is a commitment to creating safer, more productive workplaces. Their innovative solutions are transforming industrial safety. I strongly recommend MākuSafe to any organization looking to elevate their safety performance and protect their employees from the risk of injury. Their forward-thinking approach and unwavering dedication make them an invaluable partner in the quest for a safer, more secure working environment.

Bryon Snethen
(Retired) Vice President, Risk Improvement
EMC Insurance

As the Director of Client Risk Management at Henriott, an independent insurance agency and risk management consulting firm committed to helping clients mitigate and recover from critical incidents, I understand firsthand the importance of proactive safety measures. When it comes to workplace safety, especially in high-risk industries, the real priority is protecting your employees' lives and well-being. Ensuring a safer, healthier workplace is critical in today's tight labor market, and managing this risk – whether through self-insurance or traditional coverage – can make or break a company.

That's why *Safety and Wearable Technology: Impact, Applications, and Implementation in Industry* is a game changer for businesses striving for a strong safety culture and a healthy workforce. The rise of wearable safety technology is an essential evolution in risk management, providing real-time insights and predictive analytics that go beyond traditional safety measures.

The expertise of Gabriel Glynn, Mark Frederick, and Tom West is evident, having developed solutions over an eight-year period that position them as leaders in this space. Their approach demystifies the implementation of wearable technology and offers clear guidance on overcoming the cultural and operational challenges associated with these tools, ensuring that the transition to wearable safety tech isn't just a tech fad but a real strategy to protect employees and reduce costs, turning reactive strategies into proactive prevention.

For example, in "Chapter 6, Sophisticated Sensors," the authors explore how cutting-edge wearable devices equipped with environmental and biometric sensors are transforming workplace safety. These sensors monitor everything from air quality to worker movements, delivering real-time data that enables companies to react swiftly to potential hazards, reducing the risk of injuries and keeping employees safe. This proactive approach turns data into action, preventing incidents before they occur – a true revolution in safety management.

This book offers a thorough exploration of how industries can effectively integrate these innovations into their operations. It provides valuable, practical advice for organizations at every stage of their safety journey, whether they are early adopters or just beginning to explore the integration of Industry 4.0 principles.

At Henriott, we help companies both large and small manage their risks, and wearable technology is becoming one of the most effective tools for that. Whether you're a risk manager, safety professional, or a business leader looking to reduce claim frequency and severity, this book provides the roadmap you need. It's practical, actionable, and written by people who know how this tech works in the real world.

If you're serious about staying ahead of the curve in safety management and ready to take control of your insurance costs – whether through self-insurance or traditional means – *Safety and Wearable Technology: Impact, Applications, and Implementation in Industry* needs to be on your desk or iPad.

Zachary S. Finn MSM, AINS, ARM, HACCP
Director of Client Risk Management, Henriott

As a safety professional with over two decades of experience and a focus on wearable technologies, I have witnessed firsthand the evolution of safety wearables and their transformative impact on workplace safety. The history of wearables, as outlined in this book, highlights their deep-rooted presence in human society, dating back centuries to innovations like glasses and pocket watches. The authors' work in the wearable technology space and their commitment to injury prevention for almost a decade underscore the importance of the insights they share in this book.

Today, safety wearables represent the next frontier in leveraging technology to enhance workplace safety. The exponential growth in wearable technology adoption, as evidenced by the staggering market figures, underscores their increasing significance in various industries, including safety.

The historical narrative of wearables as cultural disruptors, life savers, personal entertainment devices, and always-connected tools resonates deeply with the evolution of safety wearables. Just as pocket watches revolutionized timekeeping and pacemakers extended lives, modern safety wearables hold the promise of revolutionizing workplace safety practices.

As the owner of a safety consulting company, I can foresee a future where wearable safety technology becomes ubiquitous on job sites, enhancing not only the well-being of workers but operational efficiency for safety professionals. This book serves as a valuable resource for both seasoned professionals and newcomers to the field, shedding light on the profound impact of wearables on safety practices and paving the way for a safer and more technologically advanced work environment.

I want to thank MākuSafe for allowing me to read this book and share in this experience. Here's to revolutionizing injury prevention!

Heather Chapman, MS, CSP, CHMM, CEAS III
Owner, Paradigm Safety
AIHA Total Worker Health Task Force Member
AIHA Ergonomics Committee Member
ISO 45001 Lead Auditor

I have worked in the insurance technology sector for the past 47 years, which included leading the development of two insurance agency management systems, with the last one for my own company, which was acquired by Vertafore in 1995. That system, Sagitta, is still running some of the largest insurance brokerage firms in the world. About 12 years ago, I pivoted my consulting practice from fee-for-service to equity-for-advice and, today, I am an equity adviser for nine insurtech startups.

MākuSafe is, sadly, not one of my equity insurtechs, but we became acquainted when I was running an insurtech competition for ACORD in 2019. MākuSafe won the competition that year, and I have followed their rise to prominence in the wearable safety technology space with a sort of secondhand pride ever since.

More than half the technology products that I review every year are risk-mitigating in nature, and I believe this will be the future of the insurance business. Selling a person or a business an insurance policy is necessary because we will always have unforeseen accidents to contend with, but making a person or a business safer creates a much more enduring business relationship.

While artificial intelligence is all the rage at the moment, I am a firm believer in the efficacy of machine-augmented human intelligence as a superior solution for many of our challenges. Strategically placed cameras with AI-enabled video analytics are an excellent addition to any workplace, but nothing beats a wearable device that can measure most of the environmental impacts for workers. And, perhaps the most important feature of MākuSafe wearables, the ability for workers to press a button and verbally report a safety threat or near accident situation is of paramount importance to safety-conscious employers and their insurers.

The story of MākuSafe is an invaluable one for our society!

Frank Sentner
Owner, Sentwood Consulting
Advisor & Board Member, InsurTech Hartford

Something is broken in the workplace, and there is a solution.

As the President and CEO of a portfolio of companies heavily concentrated in manufacturing, I saw firsthand that there was a problem in the workplace that was exposing employees to risk of injury, long-term bodily harm, or worse. Traditional responses to safety concerns were not resulting in substantive changes. In looking for innovative solutions, I found two important answers. The first was that meaningful actions to counter safety threats must rely on data that gives information specific to not only the work being done but the threats that are present, threats that are recognized, or the threats that are not acknowledged. The second is that utilizing that data requires a very different process of involving those closest to the work in designing and implementing systems to counter threats and to build organizational capacity. Lastly, when implemented successfully, these will lead not just to a safer workplace but to a stronger culture of engagement and can boost productivity and profitability.

Why should you read this book? You will learn about wearable safety technology that collects data to provide you with the information you need to make sound business decisions that protect your employees. You will also learn how to introduce and utilize wearable safety devices to efficiently engage those close to the work in the process of building organizational capacity. It is as important to have management read this as it is to have frontline workers read this. There must be alignment of interests, perspective, and approach to move the needle on safety and culture. The authors of the book bring years of pertinent direct involvement of developing the devices, introducing them into a variety of industrial workplaces, and working with teams at those companies on interpreting data and building solutions from the data.

Something is broken in the workplace, and there is a solution. *Safety and Wearable Technology: Impact, Applications and Implementation in Industry* will get you on the road to developing your solution.

Daniel Goldstein
CEO of Go ESOP LLC
Executive Fellow at the Rutgers Institute for the Study of Employee Ownership and Profit Sharing
Fellow at Ownership at Work
Former CEO, President, and Board Member at Folience

As a safety technologist with extensive experience designing wearable and connected work initiatives in high-risk environments, I've witnessed how wearable safety technology can enhance health and safety risk management. This book provides an essential and thorough exploration of the holistic integration of wearables into workplace systems, offering both practical insights and the foundational knowledge needed to leverage these tools effectively.

Drawing on years of hands-on experience, the authors leverage their diverse backgrounds to highlight the transformative potential of wearable technology. They provide invaluable guidance on how to implement wearables successfully, ensuring that readers are equipped to navigate the complexities of integrating this technology into risk management practices.

I highly encourage all health and safety professionals, along with their support teams, to read this book. It offers a comprehensive understanding of how wearables can enhance safety outcomes and improve worker well-being in today's connected world of work

Cam Stevens, MSc Human Factors
Founder & CEO, Pocketknife Group
Founder, Safety Innovation Academy
Trusted Guide & Coach for SafetyTech, Work Design & Digital Transformation
Australia

In today's fast-paced work environment, timely and reliable data is vital for an effective safety or continuous improvement (CI) program. Wearable technology provides these professionals with actionable insights, allowing them to proactively address root causes before they escalate. Companies using these wearables and the advanced AI analytic software are gaining significant benefits without being overwhelmed by data. For organizations with limited resources, focusing on real issues efficiently is crucial to success.

Chris H. Hill
Director, Engineering Services
Director, TAP Program
Iowa State University Center for Industrial Research and Service (CIRAS)

The future of insurance requires us to be proactive and provide solutions for risk management before claims even happen. But to provide the right solutions, we need the best data. Wearable safety products like MākuSafe are revolutionizing the industry by providing insurers and policyholders real-time information to stay ahead of the risks and protect workers.

As part of our robust loss control program at EMC Insurance, we've partnered with MākuSafe to offer policyholders this technology at preferred pricing and have seen significant improvements in workplace safety culture as a result – an incredible win for safety managers. Moreover, the data collected from these wearables has the potential to enhance our underwriting accuracy, allowing us to tailor policies more precisely to the risk profiles of our policyholders.

This book is a consequential read for anyone interested in reducing worker injury rates and is a tremendous example of how Insurtech companies are shaping the insurance industry of tomorrow.

Scott Jean
President and CEO
EMC Insurance

Wearable technology in the industrial safety space is still fairly new. While various offerings continue to better define their functionality and value, it is clear that MākuSafe has emerged as the leading platform to impact predictive outcomes.

By capturing data sets important to the health and safety of individual workers and overall facilities, and leveraging data analytics, MākuSafe empowers safety leaders to dramatically improve safety metrics and react in real time to needed modifications in job and workplace design.

Advancements in smart manufacturing and other forms of digital technology applications have impacted productivity and competitiveness in many ways. It is long overdue that a platform emerges to connect workers to the automated environments around them to greatly impact worker safety outcomes as well as productivity and efficiency. MākuSafe has found the perfect balance of data, analytics, and communication to leadership in real time, to truly impact safety in the workplace utilizing technology that is robust and cost effective.

Paul G. Jantsch
Founder and Principal, 3D Capital Investors LLC

In today's VUCA (Volatile, Uncertain, Complex, and Ambiguous) world, time has become a precious commodity for employees at all levels. The demands of the modern workplace often leave us with insufficient time to accomplish our tasks effectively. Therefore, it is essential for us to enhance our ability to identify priorities and focus areas more efficiently.

Fortunately, we now have access to innovative technologies that can significantly improve our time management. Tools such as artificial intelligence, automation, and wearables are reshaping how we work, allowing us to be both effective and efficient. At L'Oréal, we are embracing "new generation tools" that add value by conserving time and eliminating unnecessary distractions.

In the realm of health, safety, and wellbeing, these advancements have a vital role to play. Safety professionals are tasked with the important responsibility of selecting the right tools at the right time, tailoring their approach to align with the organization's cultural journey. By leveraging these technologies, we can create a safer and more productive work environment for everyone involved."

Malcolm Staves, FIChemE
Global Vice President Health & Safety
L'Oréal S.A.

Since first meeting Gabe Glynn in 2017, I have been impressed with his vision to empower workers with wearable technology. During my initial conversations on his nascent company (three employees at the time!), Gabe shared his vision: to inform workers to improve their decision-making capability; to companies to simplify through automation the collection of "actionable intelligence" (data trends); and to insurers to prospectively avoid injuries/illnesses in the workforce.

By now, that vision has matured to a successful mission. With an international impact, MākuSafe has promoted real-time data gathering that allows workers, companies, and their insurers to make data-driven decisions.

In this text, you will find the information to argue for wearable technology deployment addressing hazards of heat, forceful movement, noise, and some chemical compounds. You will also find that through the company's technology development, they have continually improved the utility of the technology and made a compelling cost/benefit analysis case for adopting the technology.

The book is a testament to the education and experiences of the company principals and the staff, as it offers practical promotion of employee engagement and metrics for transparency in exactly what is being collected and shared for data.

Read it, highlight it, then start your journey into wearable technology. There is no time to waste in pursuit of it.

Respectfully,
Steven D. Jahn, CIH, MBA, Fellow of AIHA
Past Chair, Real-Time Detection Systems Volunteer Group, American Industrial Hygiene Association

As a professional in the risk and occupational health and safety space for over 20 years, I am like many who have found myself hopeful in the benefits from wearable technology that align both strategically and tactically toward effective safeguards, communications, and controls for those that are closest to the risk: our workforce. This book is an important read for anyone wanting to learn about not only the benefits of today's wearable safety technology, but also strategic considerations for application specific to your workforce and successful implementation.

As today's world of work includes increasing complexity, length and depth within our supply chains, acknowledging how we are developing solid coupling-points between those performing the work, those impacted by potential risk exposure and driven from the desire to improve, achieving safe work in our shared world of work is taking on new meaning and urgency. Technology plays an increasingly important role in achieving this.

Better knowledge in health and safety, from Serious Incident/Fatality (SIF) intervention strategy to systems thinking and management is shedding new light on the importance of how quickly we recognize risk, determine if appropriate safeguards are in place, and enable end-to-end communication between the workers and those charged in resourcing them for safe work. I've asked these three questions below throughout my career, and I've found that leveraging the right technology, such as effective wearable solutions, enables stronger communications in each of these three areas:

1. Is work happening like we agreed/believe it's happening?
   How is the corporate end intentionally informed of how work is really happening at the working end? Has the scope of work changed? Do conflicting work instructions force employees to 'figure it out' and assume new risk?
2. What informs us/provides assurance that work is happening at acceptable levels of risk?
   What mechanism *exists for workers to easily communicate when experiencing different/new levels of risk? How are we prioritizing SIF risk related to lethal energy exposure and insufficient safeguards? How does our work system monitor/capture information for leadership to continually be informed as to the nature of work, safeguards, and resources?*
3. How are we improving?
   Do we have a systems approach that prompts continual monitoring and assessment of risk? Have we enabled those closest to the risk to capture improvement ideas and connect them to resources and leadership for improvement? Do our tools and methods for risk monitoring /assessment capture data in a manner that informs the organization (leadership) as to how work is being performed for safe outcomes, and informs leadership decisions for resources and planning?

As much as these questions have helped me in my career, what's important to recognize is the role of technology has enabled the continuity of communications within the complexity of today's work environment, while capturing data to inform the organization as to current levels of risk and continual improvement. Wearable technology plays a critical role within a healthy work system, as it connects risk information at the working end of the organization to those in a position to enable resources, change and support specific to the risk experienced by the worker. What's especially important to recognize is that the right wearable technology enables a strategic focus and intervention strategy on the type of risk that hurts people the most and creates tactical methods for monitoring and assessment at the working end, specific to the risk experienced by the worker and their job tasks.

As a safety professional who now works at a technology company focused on effective use of technology to manage risk across our supply chains, I'm grateful for the work MākuSafe is doing in the wearable technology space, as well as on the educational side toward prioritizing serious risk, strategy, and implementation specific to their clients work environments. I've coined the phrase "The right technology enables our organizations to manage risk at the speed of business", and it's refreshing to see others sharing the same drive and mission toward a better future.

Scott DeBow, CSP, ARM
Health, Safety & Environmental Director at Avetta
Author: Safety Management Systems in a Joint-Employer Environment
ASSP Board of Directors

The folks at MākuSafe haven't just penned a tutorial on wearable technology, they have introduced the interrelationship between human safety and operational health. The field of occupational safety and health is in desperate need of real-time risk data. Most workplaces are dynamic in nature and at present, safety and health risk data often comes from point sources such as assessments, inspections, and observations. This approach is analogous to checking automobile health 50+ years ago – when we had to pull over to check the oil, coolant, and air pressure. Unfortunately, there is a lot of unmanaged risk growth that happens in between these health checks. The technology outlined in the book will allow organizations to collect a real-time risk data dashboard and make educated decisions to create a safer and more successful work environment. The information herein is certain to initiate a step change in the way organizations manage their human and operational dynamic.

Peter Susca
Founder & Principal
OpX Safety / Operational Excellence LLC

Safety and Wearable Technology: Impact, Applications and Implementation in Industry is a timely resource for the occupational health and safety field. This provides an in-depth exploration into how wearable technology is actively reshaping safety management across diverse industries. With topics ranging from data-driven safety insights and IoT integration to the nuances of machine learning, the authors illuminate the powerful role technology can play in enhancing workplace safety. Beyond the technical insights, the authors explore the organizational culture and management practices essential for adopting these innovations effectively. It is a compelling read for safety professionals and industry leaders who are committed to building safer, smarter workplaces.

Nir Keren, Ph.D.
Associate Professor, A&BE, Iowa State University
Director, Occupational Safety Program, NIOSH Heartland Education and Research Center for Occupational Health and Safety
Director, VirtuTrace Laboratory for Applied Decision-Making Research in Virtual Reality

"Safety and Wearable Technology: Impact, Applications, and Implementation in Industry" is a must-read for anyone in safety and health dedicated to continuous improvement, considering the use of wearable technology as part of their safety improvement efforts, or looking for innovative ways to gain a better understanding of work as experienced by their employees. MākuSafe has been a leader in the wearable safety technology space for nearly a decade. They recognized early the transformative potential of wearable technology in safety and have been instrumental in normalizing the use of wearables to enhance workplace safety.

Wearable technology, spearheaded by organizations like MākuSafe, is here to stay and will redefine the landscape of workplace safety. The team at MākuSafe exemplifies entrepreneurship and a commitment to continuous learning. This book is a testament to their expertise and vision.

Lisa Brooks, CIE
Principal, Nexus HSE

Lisa Brooks is a certified industrial ergonomist and has been an ergonomics, S&H practitioner for over 30 years. She is known for her passion and energy and has proven leadership abilities, and distinctive technical expertise in ergonomics, human and organizational performance (HOP), and fatal and serious injury & illness prevention. She led global HSE networks at ORCHSE Strategies and The National Safety Council for 8 years after spending 23 years in industry in various global safety and ergonomics leadership roles for GE, International Paper, and Alcoa.

Glossary

Advanced manufacturing: Manufacturing that uses innovative technologies and processes to improve the design, production, and performance of products. It incorporates techniques like automation, robotics, and additive manufacturing (3D printing).

Artificial intelligence (AI): The field of computer science dedicated to creating systems that can perform tasks requiring human-like intelligence, such as learning, problem-solving, and understanding language.

ASTM E2920: A metric measuring the severity and nature of a workplace injury, used mainly to identify cases with meaningful connections to work and potential consequences that would have value for future prevention.

BBS (Behavior-Based Safety): An approach to improving workplace safety by focusing on the behaviors of employees through observing and analyzing behaviors, providing feedback, and encouraging safety practices.

Connected worker: An employee who uses digital tools and technologies, such as wearable devices, mobile apps, or IoT-enabled equipment, to access information, communicate, and receive real-time updates while on the job in order to enhance efficiency and safety.

DART (days away, restricted, or transferred): The metric used to track workplace injuries and illnesses that result in an employee being unable to work (days away), assigned to light duty (restricted), or transferred to another job as a result of the injury or illness. DART rates are used to assess the severity and impact of workplace incidents on employee productivity and can indicate a company's overall safety performance.

EHS (Environmental Health & Safety): The management and regulatory practices aimed at protecting the work environment, ensuring the health and safety of employees while complying with legal and industry standards.

Hazard: A situation or occurrence that poses a potential danger or harm to employees, property, or operations.

HOP (Human and Organizational Performance): A framework used to analyze and improve the ways people work within an organization; the focus on optimizing performance by understanding human behavior, system interactions, and organizational processes.

Human-centric technology: Technology that prioritizes user needs and experiences, ensuring that devices are intuitive, ethical, and improve quality of life.

Industry 4.0 –Digital Transformation: The ongoing Fourth Industrial Revolution, which is characterized by the integration of technologies like IoT, AI, and advanced robotics into manufacturing processes.

IoT (Internet of Things): A network of connected devices that collect and share data over the internet, enabling automation and smarter decision-making.

IIoT (Industrial Internet of Things): The application of IoT technologies specifically in industrial settings, such as manufacturing and transportation. Unlike general IoT, which connects consumer devices, IIoT connects industrial machines, sensors, and systems to monitor and optimize industrial processes, improve safety, and increase efficiency through real-time data analytics and automation.

Machine learning (ML): A subset of artificial intelligence focused on developing algorithms that enable computers to learn from and make predictions or decisions based on data without being explicitly programmed for each task.

PPE (personal protective equipment): Gear and clothing designed to protect individuals from injuries or illnesses while working in hazardous environments.

Risk: The likelihood that a hazard will result in harm as well as the severity of that harm if it occurs.

SIFs (serious injuries and fatalities) or P-SIF (potential SIFs): Incidents that result in, or have the potential to result in, severe harm or death to workers. The concept is used to prioritize and address the most critical safety risks, with the goal of preventing the most serious types of workplace accidents through targeted interventions and safety measures.

Smart factory: An interconnected and automated production environment where efficiency is enhanced through flexibility, monitoring, and adaptive control of manufacturing processes.

Index

O

P

R

S

T

V

W

Z